P9-DME-972

THE CAR STEREO
COOKBOOK

The Car Stereo Cookbook

How to Design, Choose, and Install Car Stereo Systems

Mark Rumreich

McGraw-Hill
New York • San Francisco • Washington, D.C. • Auckland • Bogotá
Caracas • Lisbon • London • Madrid • Mexico City • Milan
Montreal • New Delhi • San Juan • Singapore
Sydney • Tokyo • Toronto

Library of Congress Cataloging-in-Publication Data

Rumreich, Mark.
 Car stereo cookbook / Mark Rumreich.
   p.   cm.
 Includes bibliographical references and index.
 ISBN 0-07-058083-9
 1. Automobiles—Audio equipment.   I. Title.
TK7881.85.R85 1998
629.2'77—dc21   98-26979
                                                       CIP

McGraw-Hill

A Division of The McGraw·Hill Companies

Copyright © 1998 by The McGraw-Hill Companies, Inc. All rights reserved. Printed in the United States of America. Except as permitted under the United States Copyright Act of 1976, no part of this publication may be reproduced or distributed in any form or by any means, or stored in a data base or retrieval system, without the prior written permission of the publisher.

    12 13 14 15 DOC/DOC 0 9 8 7 6 5 4

ISBN 0-07-058083-9

The sponsoring editor for this book was Scott Grillo and the production supervisor was Pamela Pelton. It was set in Vendome by North Market Street Graphics.

Printed and bound by R. R. Donnelley & Sons Company.

McGraw-Hill books are available at special quantity discounts to use as premiums and sales promotions, or for use in corporate training programs. For more information, please write to the Director of Special Sales, McGraw-Hill, Professional Publishing, Two Penn Plaza, New York, NY 10121-2298. Or contact your local bookstore.

Information contained in this work has been obtained by The McGraw-Hill Companies, Inc. ("McGraw-Hill") from sources believed to be reliable. However, neither McGraw-Hill nor its authors guarantees the accuracy or completeness of any information published herein and neither McGraw-Hill nor its authors shall be responsible for any errors, omissions, or damages arising out of use of this information. This work is published with the understanding that McGraw-Hill and its authors are supplying information but are not attempting to render engineering or other professional services. If such services are required, the assistance of an appropriate professional should be sought.

This book is printed on recycled, acid-free paper containing a minimum of 50% recycled, de-inked fiber.

To my father, who loved music and always had the time to explain how things work.

CONTENTS

Introduction		**xix**
Chapter 1	Before You Begin	1
	Ask Yourself Some Preliminary Questions	2
	Transferability	2
	Upgradability	3
	Factory Look	3
	Understand All Your Options	3
	Understand What's in Your Car	4
	Familiarize Yourself with Available Products and Approaches	4
	Design Your System	5
	Sketch a Block Diagram	5
	Make Your Shopping List, Then Buy	6
	Read the Manuals and Draw a Complete Wiring Diagram	7
Chapter 2	Connectors, Supplies, Tools, and Techniques	9
	Connectors	10
	Soldering	10
	Crimp Connectors	11
	Quick Splice/Scotchlok Connectors	12
	Wire Nuts	14
	Fuse Taps	14
	Taps for AGC Fuse Blocks	15
	Taps for ATO/ATC Fuse Blocks	15
	Taps for MINI Fuse Blocks	16
	Installation Tips	17
	Cable Ties	18
	Specialty Tools	18
	Using a Multimeter—Wiring Harness Example	28
	Step 1: Identify the Power Wires	28
	Step 2: Identify the Ground Wire	29
	Step 3: Identify the Speaker Wires	30
Chapter 3	Speakers and Speaker Projects	33
	Imaging and You	34
	What Is Imaging?	34
	Achieving Good Imaging	36

Adding Surface-Mount Tweeters	36
Choosing Tweeters	37
Placing Tweeters	38
Connecting Tweeters	39
Upgrading Existing Speakers	41
Choosing Speakers	41
Making the Best of the Existing Enclosure	44
Installing Speakers in a New Location	45
Choosing a Location	46
Installing a Door Speaker	47
Cutting the Hole	48
Wiring It Up	50
Installing Separates	52

Chapter 4 Subwoofers and Subwoofer Projects 55

System Configuration Options	56
Conventional Speaker-Level Crossover, Subwoofer	58
Tri-Way Crossover, Subwoofer	58
Separates	59
Amps with Built-In Crossover, Subwoofer	61
Separate Crossover, Amplified Subwoofer	61
Amplified Subwoofer with Built-In Crossover	62
What to Look For in a Subwoofer Crossover	63
Choosing a Subwoofer Amplifier	65
How Much Power Do You Need?	65
The Best Amplifier/Subwoofer Configurations	66
Subwoofer Enclosure Options	69
Subwoofer Amount of Assembly Options	69
Types of Enclosures	71
Choosing a Driver	75
Power Handling	75
Sensitivity	76
Bass Response	76
Cone and Surround Material	80
Dual Voice Coil Subwoofers	80
Box Design	82
Computer Programs for Box Design	82
Step 1: Calculate f_{ob} and V_{of} from the T/S Parameters	83
Step 2: Label the Frequency Response Diagram for Your Driver	84
Step 3: Choose Your Box Size and Type	84
Step 4: Calculate the Port Tuning Frequency	93

Contents

Sealed/Ported Box Design Example   94
Bandpass Box Design Example   96
Multiple Drivers in One Box   98
Box Construction   102
    Box Materials and Shapes   102
    Assembly and Bracing   103
    Bandpass Box Construction   104
    Constructing Ports   105
    Port Design Example   106
    Speaker Terminals   108
    Make It Airtight   109
    Enclosure Damping   111
    Finishing Touches   112
Installation   113
    Bass Blocking Crossovers   113
    Trunk Sonic Isolation Check   115
    Secure Your Subwoofer   116
    Turning It On the First Time   116
    System Adjustment, or the Secret to Tight Bass 117
Bass Transducers   118

Chapter 5   Head Unit Projects   121

Why Choose an Aftermarket Head Unit?   122
Pick One That Fits   122
    Size and Mounting Method Basics   122
    Determining What Your Vehicle Can Accommodate 124
Cassette, CD, or MiniDisc?   125
Performance and Features   127
    General Features   127
    Radio Features   132
    Cassette Features   135
    CD Features   137
Installation   139
    Mechanical Mounting   139
    Electrical Wiring   146
Factory Steering Wheel Control Interfaces   151
Premium Factory Sound Systems   151
    Premium Factory Sound System Basics   152
    Choosing the Right OEM Integration Adapter 153

Chapter 6   Amplifiers and Amplifier Projects   155

Advertised Power versus Honest Power   156
How Much Power Do You Need?   158
More Power for Your Money   158
    Running 2-Ohm Loads   159
    Bridging   159
Where to Put Your Amps   161
Power and Ground Wiring   162
    Wire Gauge   162
    Power Wire Hookup   165
    Fuses   166
    Wiring for Automatic Turn-On   168
    Ground Wire Hookup   169
    Connectors   170
Speaker Wiring   171
Boosting Head Units with Four Preamp Outputs and Four
        Speaker Outputs   171
    Standard Configuration   172
    Cable-Saving Configuration   172
    Parallel Front/Rear Speakers Configuration   174
    Amplifier-Saving Configuration   174
    Cable-Saving Amplifier-Saving Configuration   174
Boosting Head Units with Two Preamp Outputs and Four
        Speaker Outputs   175
    Standard Configuration   175
    Fader-Restoring Configuration   176
    Parallel Front/Rear Speakers Configuration   177
    Amplifier-Saving Configuration   177
    Cable-Saving Amplifier-Saving Configuration   178
Boosting Head Units with No Preamp Outputs and Four
        Speaker Outputs   178
    Standard Configuration   178
    Converter-Saving Configuration   180
    Parallel Front/Rear Speakers Configuration   181
    Amplifier-Saving Configuration   181
Boosting Head Units with No Preamp Outputs and Two
        Speaker Outputs   181
    Standard Configuration   182
    Two- to Four-Speaker Upgrade Configuration   182
    Parallel Front/Rear Speakers Configuration   183
    Amplifier-Saving Two- to Four-Speaker Upgrade Configuration 183

Contents

Boosting Premium Factory Sound System Head Units   184
Premium Factory Sound System Basics   184
Do You Need to Replace Your Factory Speakers?   185
Bypassing Factory Amps   185
Line Output Converters for Bose/Ford Premium Head Units   186
System Configurations   187
Adjusting Amp Gain Settings   189
Procedure for Two-Amp Configurations   189
Procedure for Amp-Saving Configurations   190

Chapter 7   Equalizers and Equalizer Projects   191

To Equalize or Not?   192
How Many Bands Do You Need?   192
What About Equalizer Boosters?   194
Connecting Your Equalizer   194
How to Set Your Equalizer   194
Setting by Ear   195
Setting by RTA   196

Chapter 8   Biamping and Crossovers   201

Crossover Basics   202
Speaker-Level and Preamp-Level Crossovers   202
Crossover Slopes   204
Why Biamp?   206
Increased Practicality of High-Order Slopes   206
Driver Impedance Characteristic Doesn't Affect Frequency
Response   207
Tweeter Protection During Clipping of Bass   207
What to Look For in Equipment   208
Choosing a Crossover   208
Buying Amps for Biamping   210
Speakers   211
Biamping Head Units with Two Sets of Preamp Outputs   211
Separate Rear Crossover Configuration   214
Shared Crossover Configuration   214
Parallel Front/Rear Speakers Configuration   214
Full-Range Rear Configuration   214
Amp-Saving Full-Range Rear Configuration   215
Biamping Head Units with One Set of Preamp Outputs   215
Separate Rear Crossover Configuration   215
Shared Crossover Configuration   215

Parallel Front/Rear Speakers Configuration	218
Full-Range Rear Configuration	218
Amp-Saving Full-Range Rear Configuration	219
Adjusting Your System	219
Setting Tweeter High-Pass and Woofer Low-Pass Crossover Frequencies	219
Setting Woofer High-Pass Crossover Frequencies	220
Setting Woofer and Tweeter Amp Levels	221
Setting Subwoofer Crossover Frequency and Amp Level	223

Chapter 9 CD Changer Projects 225

Adding a CD Changer to Your System	226
Head Units Without CD Changer Controls	226
Aftermarket Head Units with CD Changer Controls	229
Factory Head Units with CD Changer Controls	230
Performance and Features	230
Installation	232
Choosing a Mounting Location	232
Mounting	234
Connecting	234

Chapter 10 Accessories 237

Power Line Capacitors	238
Should You Install a Power Line Cap?	238
Choosing the Right Cap	239
Installation	240
Battery Savers and Monitors	240
Premium Speaker Wire	241
Premium Patch Cables (Interconnects)	242
Portable CD Player Head Unit Adapters	244
Cassette Adapters	244
FM Transmitters	244
Antennas and Antenna Boosters	245
Embedded Windshield Antennas	246
Motorized Antennas	246
Antenna Installation	246
Antenna Boosters	247
Sound-Deadening Materials	248

Contents

Chapter 11   Battling Noise   251

   Noise Basics   252
      What Is Noise?   252
      Signal-to-Noise Ratio   252
      Signal Level and Signal-to-Noise Ratio   253
      Why High-Power Systems Are More Noise Prone   254
      Differential Versus Single-Ended Signals   255
      Ground Loops   256
      Common Ground Impedance   257
      Power Supply Noise   258
   Buying Equipment   259
      Head Units   259
      Amplifiers   260
      Crossovers and Equalizers   260
      Cables   260
      Line Drivers   261
      Master Volume Control   261
   Installation and Level Adjustment   262
      Equipment Placement   262
      Cable Routing   264
      Grounding   264
      Step-by-Step Level Adjustment   266
   Troubleshooting Noise Problems   268
      Identifying Noise Sources   268
      Shorting Plugs   268
      Ground Loop Isolators   269
      Power Line Filters   270
      Noise Sniffers   273
      Copper Shielding Tape   275
      Step-by-Step Troubleshooting Procedure   276

Recommended Mail-Order Suppliers   279

Recommended Reading   280

Bibliography   283

Index   287

PREFACE

While browsing a popular car audio forum on the Internet, I was struck by how many good questions people were asking and how many well-intended but incorrect answers were given. A trip to the bookstore later that week didn't turn up much useful information on car stereo either—surprising considering the popularity of the subject. That's when I decided to start researching and start writing.

Whether you realize it or not, there's a war being waged over the radio-sized hole in your dashboard. Automobile manufacturers want you to buy a factory sound system and will do everything in their power to sell you one. Face it, vehicle manufacturers have strategic advantages. They can style the radio to match the dash and tailor speaker placement and equalization to match the acoustics of each vehicle (few go to the trouble, however). They're not above some dirty tricks, either. Integrating climate controls into the radio, nonstandard equipment mounting sizes, proprietary CD changer protocols, and hard-to-interface-with premium factory sound systems have made aftermarket installation more challenging.

But the aftermarket is up to the challenge, with high-performance solutions to almost any problem. The aftermarket itself is far from being a uniform front in the car stereo war, however. Numerous aftermarket companies are vying for your hard-earned dollar too.

It's hard to get a straight story on what you really need to get the performance you want. And the rules are constantly changing. That's where this book comes in. It provides the right answers to the growing list of questions—so you can win the war!

—MARK RUMREICH

ACKNOWLEDGMENT

This book would not have been possible without considerable help from a lot of people, including my wife and family. Extra thanks to Alan Hoover for his thoughtful and painstaking review of the manuscript.

—MARK RUMREICH

INTRODUCTION

The Car Stereo Cookbook provides a comprehensive and well-organized guide to designing, choosing, and installing car stereo systems. It's geared toward the technically oriented do-it-yourselfer who is interested in car stereo. It assumes some familiarity with car stereos, but you don't need to have a degree in electronics or have installed a stereo before.

The Cookbook Approach

Unlike other car stereo books and magazines, which describe complete installations for specific vehicles, *The Car Stereo Cookbook* offers a more a la carte approach. Chapters are organized by the type of project, such as speaker projects or amplifier projects. Projects may be used individually or combined to create a more advanced installation.

The Car Stereo Cookbook covers an entire spectrum of projects from subwoofers to accessories and from basic to advanced. It also presents a range of cost options. Step-by-step instructions are provided for difficult procedures and just the right amount of explanation is given to help you make smart choices.

Why Do It Yourself?

There are many reasons to do it yourself—saving money, making sure it gets done right, the freedom to select exactly the products you want, and the satisfaction of knowing that you did the job. This book will help you meet all these goals.

How This Book Is Organized

The first two chapters provide useful preparatory information. Chapter 1, "Before You Begin," describes how to plan your system, from considering upgradability, transferability, and maintaining a factory appearance to drawing a wiring diagram. Chapter 2, "Connectors, Supplies, Tools, and Techniques," covers popular methods of connecting wires, describes the basic specialty tools useful for car stereo work, and explains how to use a multimeter.

The heart of the book is composed of the project chapters. These cover the following types of projects:

- Speakers and speaker projects
- Subwoofers and subwoofer projects
- Head unit projects
- Amplifiers and amplifier projects
- Equalizers and equalizer projects
- Biamping and crossovers
- CD changer projects
- Accessories

Chapter 3, "Speakers and Speaker Projects," starts by explaining what good imaging is all about and how to achieve it. Projects range from adding surface-mount tweeter modules to installing a component speaker system.

Chapter 4, "Subwoofers and Subwoofer Projects," is by far the largest chapter. This is for good reason, since subwoofers play such an important role yet are frequently misunderstood. There are clear explanations of how to choose the right crossover, amplifier, driver, and box. For the more adventurous, box design and construction for sealed, ported, and band-pass boxes are covered.

Chapter 5, "Head Unit Projects," explains what to look for in head units and how to install them. It also includes a section on upgrading head units in premium factory sound systems such as Delco/Bose or Ford/JBL.

Chapter 6, "Amplifiers and Amplifier Projects," explores how much amp power you need and how to get more power for your money. Power, ground, and speaker wiring are covered in detail. Projects cover boosting head units of all types and include procedures for properly setting amplifier gains.

Chapter 7, "Equalizers and Equalizer Projects," explains the differences between graphic and parametric equalizers and how to choose the right number of bands for your application. Installing and setting up an equalizer, both with and without a real-time analyzer, are explained in detail.

Chapter 8, "Biamping and Crossovers," details the advantages of biamping, covers the basics of crossovers, and tells you what to look for in equipment. Projects cover biamping with head units of all types and include procedures for adjusting the system.

Chapter 9, "CD Changer Projects," explains the different methods of interfacing a changer to a system as well as what to look for in a changer. Projects cover installation of changers with head units of all types.

Chapter 10, "Accessories," separates science from snake oil when it comes to power line caps, premium speaker wire, premium patch cables, and sound-deadening mats. Battery savers, portable CD player adapters, and antennas are also discussed.

The last chapter is Chap. 11, "Battling Noise." This chapter explains the common causes of automotive noise problems and how to fix them (or better yet, avoid them).

CHAPTER **1**

Before You Begin

This chapter is intended to get you thinking about some of the less obvious, but still important, factors in planning your system.

Ask Yourself Some Preliminary Questions

Here are a few questions you need to ask yourself *before* you start to design your system:

- How long am I going to keep this car? Will I sell the stereo with it?
- Is this a one-shot deal, or might I want to upgrade later?
- Do I want a factory look?

Transferability

It's hard to think ahead to the time when you'll be selling (or possibly junking) your car. One day it will happen, and on that day you'll either have to remove your upgraded stereo (and possibly restore the original one) or sell it with the car.

Putting in a really awesome sound system is almost like installing a built-in swimming pool in your backyard. It's worth every cent to you, but it's hard to get your money back when it comes time to sell. At least with a stereo, you can take it with you. Unfortunately, you may leave behind gaping holes in the dash, doors, and deck.

This implies a few things. First, don't drill or cut any holes that you can't easily repair, conceal, or accept, unless you're willing to leave what's in them. Cutting the dashboard should be avoided if at all possible. Second, NEVER cut off factory wiring harness connectors. Either splice into the wires or spend a few extra bucks and buy wiring harness adapters. Third, try to choose equipment that is easily transferable to another vehicle. This means things such as buying a subwoofer box that will fit in your next car or truck as well as your current one.

It makes more sense to sell less dramatic upgrades with the car. Things like upgraded speakers or a well-chosen head unit can add value to a vehicle at selling time.

Elaborate stereos should be easily uninstallable to an acceptable condition.

Upgradability

You should also consider what type of improvements you may make to your system in the future. A few extra dollars now may save you hundreds later, as well as considerable time. Some examples:

- Buying a head unit with two or three sets of preamp outputs will expand your options if you later decide to add amplifiers, a subwoofer, or electronic crossover. If you think you might add a CD changer down the road, CD changer controls are a must.
- An amp that has speaker-level and preamp-level inputs and that is bridgeable stands a much better chance of being useful in the next incarnation of your system.
- If you think you might want to biamp later, choose component speakers or biampable coax drivers when upgrading speakers now. These provide excellent performance with their passive crossovers (normally included) and are ideal for biamping. See Chap. 8 for details.

Factory Look

Maintaining a factory appearance is important to many people for aesthetic as well as theft deterrent reasons. Others prefer to showcase their installations and protect their investment with an alarm system.

Having good sound while maintaining a factory appearance is easier than ever before. Today's factory head units can provide excellent performance and can be connected to external amplifiers if you need more power. Factory speakers can be replaced with better aftermarket drivers, and you can conceal a powered subwoofer in the trunk to provide a rock-solid bottom end. Add-on CD changers that will work with any factory head unit are readily available.

Even if you decide to replace the factory head unit, aftermarket models designed to match the factory styling are available for many vehicle makes. These offer CD changer controls and even plug into the factory wiring harness.

Understand All Your Options

It's easy to spend more money than you need to on a car stereo system. You can avoid making this mistake by understanding all your options before you design your system. This means understanding what's in your car now, what products are out there, and what approaches will meet your needs.

Understand What's in Your Car

The first step in understanding all your options is investigating the stereo you have in your car now. This is particularly important if you plan to upgrade rather than start from scratch.

If you bought your vehicle new from the factory, you should have a pretty good idea of what you've got. But if you bought from a used car dealer, you may be in for a surprise (pleasant or otherwise). The previous owner may have upgraded the speakers or added an amp under the dash. On the other hand, maybe the wiring harness was butchered or the cassette deck has an intermittent problem.

You can tell a lot just by listening, but it's a good idea to at least take a look at the back of a speaker in the trunk for evidence implying an upgrade of some sort. You may have a premium factory sound system, in which case amplified speakers are probably involved. This can be a blessing or curse, depending on what you want to do.

Familiarize Yourself with Available Products and Approaches

The more familiar you are with what products are out there and how to use them, the better you'll be able to design a system that sounds great and is affordable. Study catalogs and visit your local car stereo shop. Check out the latest products—some could be just the answer you need. And use this book to familiarize yourself with both common and not-so-common car stereo solutions and how to squeeze the most out of each component you buy.

Some of the best car stereo solutions are often the ones that aren't obvious. A few good examples:

■ Using amplifier-saving configurations for boosting head units. These can save you the cost of an amplifier with no appreciable loss in performance.

■ Adding surface-mount (or flush-mount) tweeters rather than upgrading front factory speakers. This can save money and provide better high-frequency reproduction and stereo imaging.

■ Adding a subwoofer to fix a system that isn't loud enough. You'll be able to turn down the bass control to your main speakers and use the subwoofer to provide a rock-solid bottom end. By transferring the burden of bass to the subwoofer, you enable the rest of your speakers to play louder with less distortion. This frequently eliminates the need to boost the head unit with an amplifier.

Design Your System

Whether you're starting from scratch or upgrading, the other chapters in this book should give you the basic information you need to design your system. If you're thinking of upgrading, Table 1-1 can help point you in the right direction.

Sketch a Block Diagram

As a part of your system design process, you should sketch a simple block diagram of the system you intend to create. The purpose of this diagram

TABLE 1-1

Upgrade Solutions
to Common
Complaints

Factor	Possible Solutions	Chapter
Not loud enough	Replace head unit with high-power model	5
	Boost head unit with external amp	6
	Add a subwoofer	4
Poor treble response	Upgrade speakers	3
	Add surface-mount tweeters	3
Want to play CDs	Install a CD head unit	5
	Add a CD changer	9
	Use an outboard CD player	10
Weak bass	Add a subwoofer	4
Boomy/muddy bass	Add a subwoofer	4
	Install an equalizer	7
Poor imaging/staging	Imaging upgrade	3

is to make sure you haven't forgotten some key piece of equipment (like an amplifier, or maybe a line output converter to convert speaker-level signals to preamp level). The diagram doesn't need to be anything fancy. It should show each piece of equipment in the system as well as the basic connections. Your diagram might look something like the one in Fig. 1-1.

If you're putting in multiple high-power amplifiers, you should include power distribution hardware (such as fuse blocks) in your diagram.

Make Your Shopping List, Then Buy

Once you've finished a simple block diagram of your system, make your shopping list. This will reduce the chances of buying the wrong stuff and having to return it and/or needing to make multiple trips to the store during the middle of your installation.

Your shopping list should include everything on your block diagram, plus tools and supplies you'll need for the job. See Chap. 2 for advice on specialty tools you may want to buy. Don't forget:

■ Speaker wire (not needed for most simple head unit installations)

■ Power, ground, and remote turn-on wire (for an outboard amp)

■ Patch cords (for an outboard amp)

■ Wiring harness adapters

Figure 1-1
A simple block diagram.

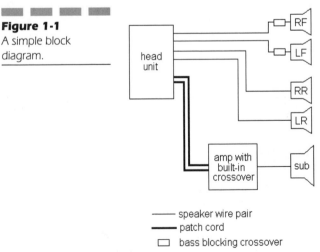

- Connectors, heat-shrink tubing, and wire ties
- Solder

As you're deciding which models of equipment to buy, visit your local car stereo shop. They can demonstrate what specific head units, speakers, and subwoofers actually sound like and can calibrate you on how much power you need. Be sure to choose a shop that has a reputation for a knowledgeable and friendly staff and that carries quality equipment.

You'll also want to check out the latest Crutchfield catalog for the newest products and a wealth of information. Its Easy-Fit application guide lists head unit and speaker sizes for most vehicles. Product performance and feature charts show you what to look for and simplify the comparison process. Crutchfield maintains an extensive database, and their well-trained staff will alert you to special adapters and tools you'll need for your project. They also provide detailed, vehicle-specific Master-Sheet installation instructions with head unit or speaker purchases.

See the list of recommended companies at the end of the book for other mail-order companies.

Read the Manuals and Draw a Complete Wiring Diagram

Once you have all your equipment in hand, it's time to read the pile of literature that came with it. As exciting as it is to dive right in and start installing, it's smarter to read the owner's manuals first. You may discover a fundamental problem in your plans due to a bad assumption about a particular piece of equipment.

A good example is head unit faders. In almost all models with one set of preamp outputs, these are controlled by the fader—but in a few they are not. If you were counting on fader control of your preamp outputs, this could throw a wrench into your plans.

After you've read the owner's manuals, it's a good idea to draw a more complete wiring diagram for your system. If all you're doing is upgrading factory speakers or replacing a head unit, then you can skip this step. If you're doing a complex multiamp system, then a detailed wiring diagram can go a long way toward avoiding hooking things up wrong. You should show every wire with its color and function indicated. Your diagram might look like the one in Fig. 1-2.

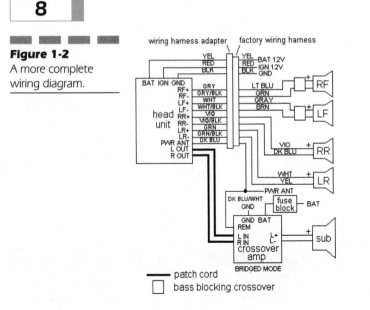

Figure 1-2
A more complete
wiring diagram.

Drawing a complete wiring diagram can be the least fun part of preparing for your installation. But the payoff comes when you wire up the system quickly and without errors. Another payoff comes down the road when you need to repair the system or if you decide to upgrade.

Now you're ready for installation!

CHAPTER **2**

Connectors, Supplies, Tools, and Techniques

Discovering you don't have the right tool in the middle of a job can be frustrating. In many cases people are not aware that some of the most useful tools even exist, and continue to "drive screws with a hammer."

This section is intended to make you aware of the most useful of the car stereo specialty tools and supplies. Most of the tools only cost a few dollars, and can be used for other projects around the house. The time to buy tools and supplies is *before* you start your car stereo project.

Connectors

Talking about connectors before tools may seem backward, but you should think about connectors first since they may affect what tools you need.

Most car stereo projects involve making lots of connections. Table 2-1 shows the four commonly used methods of making connections—soldering, crimp connectors, Scotchlok™ connectors, and wire nuts.

Soldering

Soldering produces the best electrical connection by far. It provides a low-impedance, corrosion-resistant, mechanically strong connection. Unfortunately, soldered connections are also the most time-consuming, both in soldering and in insulating. And a hot soldering iron always seems to find its way to the upholstery.

TABLE 2-1

Common Connection Methods

Method	Advantages	Disadvantages
Soldering	Best electrical connection	Time consuming, must insulate
⟨BEST⟩ Crimp connectors	Good electrical connection, self-insulating, a type for any connection	Can corrode or pull apart if improperly crimped
Quick splice/ Scotchlok	Good electrical connection, self-insulating, easiest way to make a splice	Can corrode, bulky, for splices only
Wire nuts	No special tools needed	Easily become untwisted— not recommended for car stereo

Soldering Tips

Soldering requires skill and patience. To get good results, follow these tips:

- Use only 60/40 (60 percent tin, 40 percent lead) rosin-core solder.
- Make sure the connection to be soldered is free of dirt and corrosion.
- Periodically "tin" the tip of the soldering iron by coating it with solder and brushing off the excess with a cloth until it is smooth and silvery.
- Always heat the connection, not the solder.
- Use the side of the soldering iron tip near the point, not the point itself, to heat the connection. This allows more heat to flow from the tip into the connection.
- A proper solder joint is smooth and silvery, not rough and gray. Rough and gray joints are called *cold* joints because not enough heat was used to melt the solder. Always resolder cold joints.

The traditional way to insulate a soldered connection is with electrical tape, but the adhesive used with most tapes gets gummy after a few years and the tape is likely to slide off. A better solution is heat-shrink tubing. Heat-shrink tubing must be slid onto the wire before soldering; then it is slid over the soldered connection and heated with a butane lighter or heat gun until it shrinks to conform to the connection.

Crimp Connectors

Crimp connectors are the most popular connectors for autosound. There are crimp connectors available to make almost any type of connection imaginable for any type of wire (Fig. 2-1).

Disconnectable terminals are available for ease of disassembly (for repair, for example). To use, simply strip ¼ inch of insulation from the

Figure 2-1
A world of crimp connectors: (left to right) spade, ring, male disconnect, female disconnect, bullet plug, bullet receptacle, butt splice, closed end.

end of the wire, insert into the crimp terminal, and crimp with a terminal crimping tool. Try to pull the wire out of the terminal to verify a good connection.

The key to success with crimp terminals is to have the right size connector for the gauge of wire. If you try to use a connector that is too large for the wire gauge, it will not make a good connection and is likely to slide off. Terminals are generally color coded according to the industry standard shown in Table 2-2. If you don't know the wire gauge, use a crimper/stripper to figure it out by stripping some insulation, or just choose the smallest crimp connector that fits the wire. Be sure to use the pull test for each crimp.

A trick you can use when a wire is too small for a connector is to first fortify the wire by twisting additional strands around it (Fig. 2-2). This is useful for super-fine wires or for making a butt connection between two wires of different gauges.

Quick Splice/Scotchlok Connectors

Scotchlok connectors are the best way to splice into an existing wire without having to cut it. This connector opens like a clamshell and has channels for the existing wire and the wire to be spliced into it (Fig. 2-3). After the wires are put in their channels, the clamshell is closed with a pair of pliers. Since these connectors cut into the insulation, the insulation should not be stripped. Power and ground wires are a good example of common uses of these connectors. Scotchlok connectors can also be used to connect an aftermarket head unit to a factory wiring harness without cutting off the end of the harness. This is useful if a wiring harness adapter isn't readily available and simplifies things if you ever decide to reinstall the factory radio.

There are two keys to success with Scotchlok connectors:

TABLE 2-2

Color Code for
Crimp Connectors

Wire Gauge	Standard Crimp Connector Color
22–18	Red
16–14	Blue
12–10	Yellow

Figure 2-2
(*a*) Fortification trick—before; (*b*) fortification trick—after.

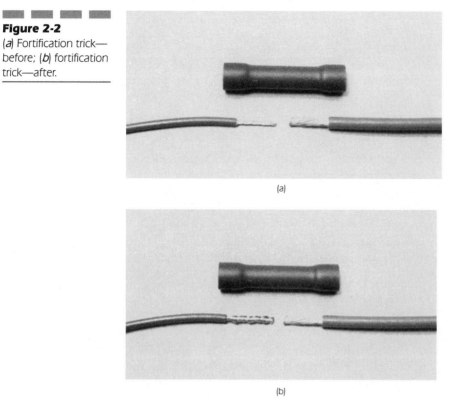

(a)

(b)

- Be sure to use the right size connector for the wire. Too large a connector will result in a poor (or no) connection. Too small a connector will cut through many of the wire strands, weakening the wire.
- Be very careful when centering all the wires within their channels in the connector to prevent cut wires or bad connections.

Scotchlok connectors are usually color coded as shown in Table 2-3. Most connectors specify the same wire gauge range for both wires, but

Figure 2-3
Scotchlok connector.

TABLE 2-3

Color Code for
Scotchlok
Connectors

Wire Gauge	Standard Color
22–18	Red
18–14	Blue
12–10	Yellow

some specify one range for the run wire and another for the tap wire. This lets you splice into a thick wire with a thin wire.

If you're unsure about the wire gauge or connector, compare the size of the gap in the metal teeth of the Scotchlok connector against the wire's *conductor* diameter. The gap should be roughly one-third to one-half the conductor diameter.

Wire Nuts

Wire nuts (Fig. 2-4) are the standard connector for home wiring, but are not intended for car stereo. They can easily untwist and cause a short or open circuit. This may result in damaged equipment or equipment that simply fails to operate.

Fuse Taps

When you need a source of power for a new accessory, you can splice into one of the existing power wires. However, a better alternative is to use a fuse tap. Fuse taps provide an easy way to tap into an automotive fuse

Figure 2-4
Wire nuts.

block. They come in a number of styles, depending on the type of fuses your car uses.

Taps for AGC Fuse Blocks

For AGC (glass tube) fuseholders, use AGC fuse taps (Fig. 2-5). Temporarily remove an existing fuse, insert the fuse tap, then reinsert the fuse. The fuse tap provides a ¼-inch male solderless terminal for the accessory power wire to connect to.

Taps for ATO/ATC Fuse Blocks

For ATO/ATC (blade) fuseholders, you have more styles to choose from.

The *clip-style fuse tap* (Fig. 2-6) is the least expensive option. You temporarily remove an existing fuse, clip the tap onto the fuse, then reinsert the fuse. The fuse tap provides a ¼-inch male solderless terminal for the accessory power wire to connect to.

Tip: Clip-style fuse taps that wrap around the fuse terminal can mechanically overstress the terminals in the fuse block. This can cause nuisance blows due to excessive heat produced by a loose-fitting fuse. A "through the hole" design (shown in Fig. 2-6) avoids this problem.

The *lead-wire fuse* (Fig. 2-7) is actually an ATO/ATC fuse with a 7-inch fully insulated lead wire coming out the back. One advantage of this approach is that there is no risk of mechanically overstressing the terminals in the fuse block. Another is that the lead wire can be bent fully back, allowing the fuse box cover to fit back on. A disadvantage of this approach is that since the fuse itself is part of the assembly, you'll need to buy a lead-wire fuse with the proper current rating. If the fuse ever blows, you'll need to replace the entire assembly and redo the accessory connection. If you choose this approach, be sure to order a spare. Lead-wire fuses are available from MCM Electronics.

The *Add-A-Circuit™ fuse holder* by Littelfuse (Fig. 2-8) plugs into the fuse block and has two fuse holder slots built in. The first slot is for the origi-

Figure 2-5
AGC fuse tap.
(*Courtesy of MCM Electronics.*)

Figure 2-6
ATO/ATC clip-style
fuse tap. (*Courtesy of
MCM Electronics.*)

nal circuit and the second is for the new accessory circuit. A fully insulated lead wire is provided for the new accessory circuit. This ingenious design eliminates having to install an in-line fuse holder for the accessory. Its only drawback is its size—you may have difficulty fitting the fuse box cover back on.

Taps for MINI Fuse Blocks

Many newer vehicles use MINI fuses in their fuse blocks. These are miniature blade-type fuses, similar to ATO/ATC fuses, but smaller. For this type of fuse, you can use MINI clip-style fuse taps (Fig. 2-9).

Littelfuse also makes an Add-A-Circuit fuse holder for MINI fuses.

Figure 2-7
ATO/ATC lead-wire
fuse. (*Courtesy of MCM
Electronics.*)

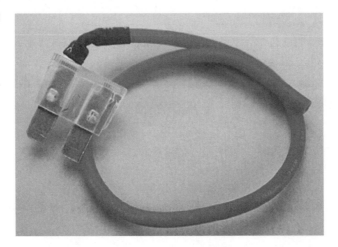

Figure 2-8
ATO/ATC Add-A-
Circuit fuse holder.
(*Courtesy of MCM
Electronics.*)

Installation Tips

With any type of fuse tap, you'll need to identify which side of the factory fuse is connected to the battery and which side is connected to the load. You'll need to use a multimeter for this. Remove the factory fuse, turn the ignition on, and connect the black lead of the multimeter to the vehicle's chassis ground. Touch the red test lead of the multimeter to each of the fuse terminals. The terminal that shows approximately +12 volts is the battery side of the fuse. The terminal that shows approximately 0 volts is the load side.

Now you'll need to decide whether to locate the fuse tap on the battery side or the load side of the existing fuse. For the Add-A-Circuit fuse holder, the lead wire should be located on the load side. For the other types of fuse taps, it's a matter of choice. Using the battery side requires that you add a fuse to the accessory power wire as close to the factory fuse block as possible, for safety reasons. The advantage of this approach is that the accessory cannot cause the factory fuse to blow and the factory circuit cannot cause the accessory fuse to blow. Locating the tap on the load side of the factory fuse is the safest and best approach if the accessory draws only a small amount of current (less than 1 ampere).

Figure 2-9
MINI clip-style fuse tap.
(*Courtesy of MCM
Electronics.*)

When adding a fuse tap, tap a fuse on a circuit labeled IGNITION if you want accessory power only when the ignition switch is on. If you want accessory power all the time, choose a fuse on a circuit labeled BATTERY.

Cable Ties

Cable ties (also called wire ties or tie wraps) are so handy, they deserve their own section. They are molded of high-tensile-strength nylon and are used for bundling together wires and cables. The flat, flexible tail is inserted through a ratcheting slot in the head to create a self-locking, permanent loop (Fig. 2-10).

Cable ties come in a variety of sizes—4 and 8 inches are the two most popular lengths for car stereo work (Fig. 2-11). They are most commonly white or black, but virtually every color is available, including neon colors for the fashion-conscious installer.

Uses for cable ties include:

- Bundling wiring together for a neat installation.
- Routing wiring out of sight and out of harm's way by cable-tying wiring to objects along the desired path.
- Securing crossovers or small modules under the dash, inside doors, or in the trunk.
- Marking wires—use different numbers of or different colors of cable ties to identify identical-looking wires. You can also buy special identification cable ties with a built-in writeable tab.

Specialty Tools

In addition to the usual screwdrivers, wire cutters, pliers, and wrenches, there are a number of tools worth investing in, even if you only install a

Figure 2-10
How a cable tie works.

Figure 2-11
World of cable ties.

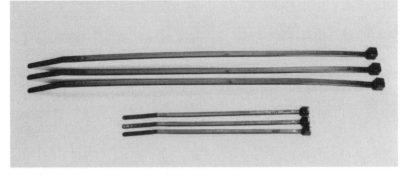

stereo once. The following tools are especially useful for working on car stereos, although most of them are also useful for other jobs. The list in Table 2-4 includes all my personal favorites.

A **crimper/stripper** (Fig. 2-12) is a necessity if you plan to use crimp connectors. These are often packaged with an assortment of terminals, which is a good way to start. Cheap models often do a poor job of crimping, which can result in bad connections over time. Most crimper/strippers cover the range of 22-gauge to 10-gauge wire terminals—more than adequate for most projects. For 8-gauge wire or lower, you'll need to use an expensive impact or lever-style crimper or switch to set-screw-type terminals.

A **soldering iron** can be purchased for as little as $5, and it will do a fine job. More expensive irons offer features such as a built-in stand, interchangeable tips, and temperature control. Most soldering irons are 120-volt models—they plug into a wall outlet (Fig. 2-13). There are two types of soldering irons of particular interest for car stereo work—the 12-volt iron and the butane iron.

The 12-volt iron (Fig. 2-14) plugs into your cigarette lighter socket. It lets you solder in your car without having to use an extension cord.

Butane soldering irons (Fig. 2-15) start at about $30. They refill using standard butane lighter fuel. Butane models are nice because they can solder where there is no power of any kind and they don't have a cord to get in your way. They also can be used with optional tips such as a hot knife (for cutting plastic) and a hot blower (perfect for shrinking heat-shrink tubing or tape). Optional tips cost about $10 each. A butane iron is really nice if you can afford one.

A **ratcheting offset screwdriver** (Fig. 2-16) is often the only good way to access screws in tight places. Dashboard work (such as installing a new head unit) is the most common situation requiring this tool. Replacing speakers in doors is another job that may require an offset screwdriver.

TABLE 2-4

Useful Tools for Car
Stereo Work

Tool	Cost	Comments
Crimper/stripper	$3—10	Must have for crimp connectors
Soldering iron	$5 and up	Must have if you plan to solder
Ratcheting offset screwdriver	$12—15	Must have for some jobs
Hollow-shaft nutdrivers	$10—25	Luxury
Torx (star drive) driver set	$5—20	Must have for some jobs
SnakeLight™	$20	Luxury
24-inch claw pick-up tool	$3	Very handy
Utility hacksaw	$4	Very handy
Utility knife or X-Acto knife	$3	Very handy
Black paint pen	$3	Great for minor chips and scratches
Window crank clip remover/installer	$4—6	Luxury
DIN removal tools	$0—3	Needed to remove DIN-E head units
Test CD	$10—25	Simplifies tweaking your system
Multimeter	$15 and up	Recommended

Skewdriver™ and Skewdriver Jr.™ offset screwdrivers perform the same job, but with a screwdriver handle instead of a ratchet handle.

Hollow-shaft nutdrivers (Fig. 2-17) are useful for removing or installing hex-head screws and hex nuts. The hollow shaft makes them ideal for the shallow nuts on head unit control shafts (such as the volume control). A socket set with a screwdriver handle adapter makes a good substitute for most applications. Metric is standard today on new cars and equipment, but if you're working on older cars, you'll want English sizes too.

Figure 2-12
Crimper/stripper.
(*Courtesy of MCM Electronics.*)

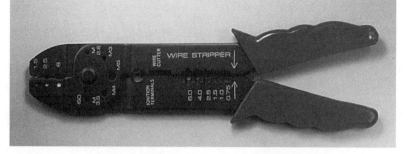

Figure 2-13
Soldering iron—
120 volts. (*Courtesy of MCM Electronics.*)

Figure 2-14
Soldering iron—
12 volts. (*Courtesy of MCM Electronics.*)

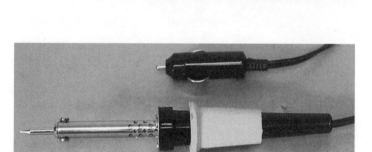

Figure 2-15
Soldering iron—
butane. (*Courtesy of MCM Electronics.*)

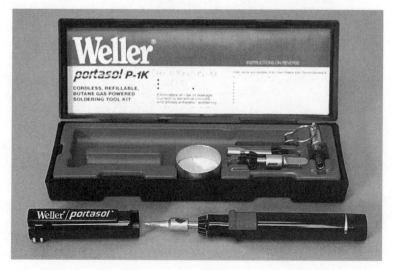

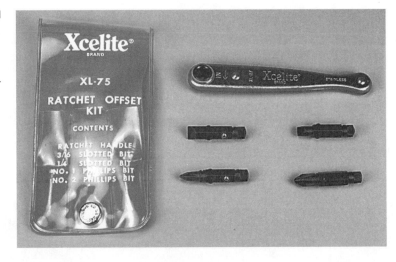

Torx (star drive) screws are widely used in cars for dashboards and speaker assemblies. They are also used for many appliances and electronic items, so **Torx drivers** (Fig. 2-18) are good to have around the house. Unlike with flat or Phillips screws, you must have exactly the right size driver for the screw. The most common sizes are T10, T15, T20, T25, T27, T30, and T40. T50 is needed for some seat belt bolts.

To economize, you can buy a Torx key set for as little as $5. For a little more, you can buy a set of Torx screwdrivers. Yet another option is a set of Torx bits that can be used in a cordless screwdriver. Most ratcheting off-set screwdrivers and Skewdrivers come with a set of small Torx bits, so you can kill two birds with one stone by buying one of these.

Figure 2-17
Hollow-shaft nut-
drivers. (*Courtesy of
MCM Electronics.*)

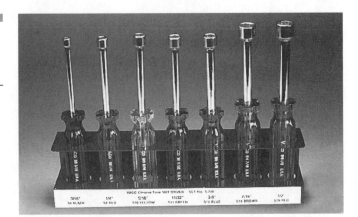

Figure 2-12
Crimper/stripper.
(*Courtesy of MCM
Electronics.*)

Figure 2-13
Soldering iron—
120 volts. (*Courtesy of
MCM Electronics.*)

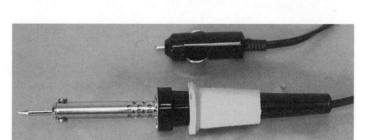

Figure 2-14
Soldering iron—
12 volts. (*Courtesy of
MCM Electronics.*)

Figure 2-15
Soldering iron—
butane. (*Courtesy of
MCM Electronics.*)

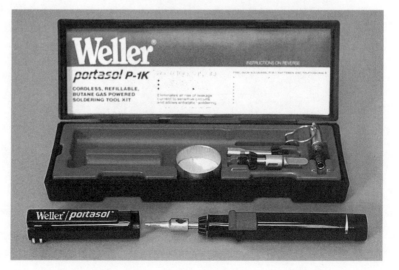

Figure 2-16
Ratcheting offset
screwdriver. (*Courtesy
of MCM Electronics.*)

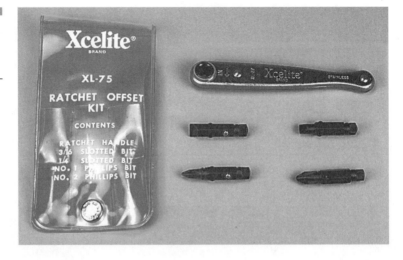

Torx (star drive) screws are widely used in cars for dashboards and speaker assemblies. They are also used for many appliances and electronic items, so **Torx drivers** (Fig. 2-18) are good to have around the house. Unlike with flat or Phillips screws, you must have exactly the right size driver for the screw. The most common sizes are T10, T15, T20, T25, T27, T30, and T40. T50 is needed for some seat belt bolts.

To economize, you can buy a Torx key set for as little as $5. For a little more, you can buy a set of Torx screwdrivers. Yet another option is a set of Torx bits that can be used in a cordless screwdriver. Most ratcheting offset screwdrivers and Skewdrivers come with a set of small Torx bits, so you can kill two birds with one stone by buying one of these.

Figure 2-17
Hollow-shaft nut-
drivers. (*Courtesy of
MCM Electronics.*)

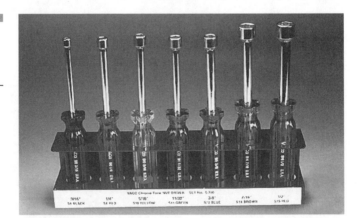

Figure 2-18
Torx driver set.
(*Courtesy of Stanley.*)

The **SnakeLight** (Fig. 2-19) is a flexible flashlight particularly well suited to working under dashboards. Its flexible shape lets it fit into tight spots as well as hold itself in place. There are numerous clones available. There are also many other specialty flashlights that have clamps, head-straps, or magnets to give you two hands to work with. A no-hands flash-

Figure 2-19
SnakeLight. (*Courtesy of Black & Decker.*)

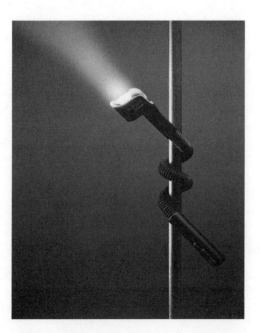

Figure 2-20
24-inch claw pick-up tool. (*Courtesy of MCM Electronics.*)

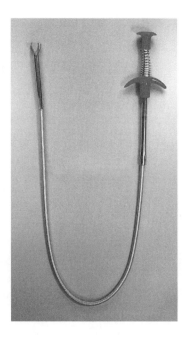

light is handy to have around the house as well. Be sure to get something bright—a halogen, xenon, or krypton bulb is a real plus.

A **claw pick-up tool** (Fig. 2-20) is indispensable for retrieving small parts dropped inside a car door or behind the dashboard. Another great use for this tool is in starting screws in hard-to-reach spots. It's also very handy for things other than car stereo work, like retrieving a key dropped through a crack in your deck.

A **utility hacksaw** (Fig. 2-21) is the best tool I have found for cutting small holes in dashboards, door panels, or rear decks. It's slow compared to a jigsaw, but doesn't require a flat working surface to rest the saw on. It also reduces the risk of damage from scratching the surrounding area.

Figure 2-21
Utility hacksaw. (*Courtesy of Stanley.*)

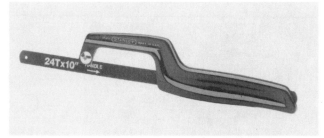

Figure 2-22
Utility knife. (*Courtesy of Stanley.*)

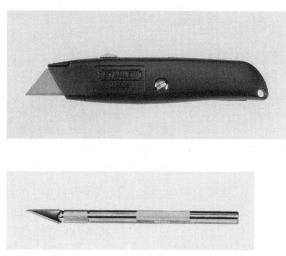

Figure 2-23
Precision knife. (*Courtesy of Stanley.*)

A **utility knife** (Fig. 2-22) is useful for cutting and trimming wood, plastic, and upholstery. A precision knife (Fig. 2-23) performs the same jobs, but provides better control. A variety of blades are available for precision knives, but the standard triangular pointed blade is best for most jobs.

A **black paint pen** (Fig. 2-24) looks like a magic marker, but it dispenses fast-drying black paint instead of ink. It does a great job of concealing minor chips and scratches and slightly damaged painted screw heads, inevitable during even careful installations. It's perfect for painting the thin metal lip visible along the edge of some DIN mounting sleeves. Black permanent markers work in a pinch, but don't cover evenly and, in some light, take on a reddish cast. Once you use a paint pen, you'll never want to go back. Paint pens are also available in various other colors including white, silver, and gold.

When installing door speakers, you will need to remove the window crank in order to remove the door panel. Some window cranks are held on with a screw. Most are held in place by a spring clip. For the most part, a piece of bent coat hanger will do the job of removing the spring clips

Figure 2-24
Black paint pen. (*Courtesy of Sanford.*)

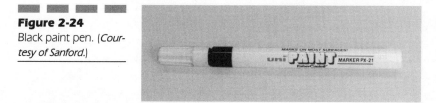

Figure 2-25
Window crank clip
remover/installer.
(*Courtesy of MCM
Electronics.*)

(just bend a tiny hook in the end). A **window crank clip remover/
installer** tool (Fig. 2-25) helps prevent the clips from flying into space.

DIN removal tools (Fig. 2-26) are just two U-shaped pieces of solid
wire. You'll need them to pull a Euro DIN head unit out of a dash. These
are often included with a head unit purchase, but are also sold separately.
This is yet another tool that you can make out of bent coat hangers to
save a few bucks.

Tweaking your system for the best sound is easier with a good test CD
(Fig. 2-27). There are lots of **test CDs** to choose from—which is best
depends on what your project is and whether you have a sound pressure
level meter to take advantage of some of the test tones. Some test discs are
made for a specific purpose such as setting amplifier levels, "system con-
ditioning," or preparing for auto sound competition.

A **multimeter** (Fig. 2-28) is the single most important piece of test equipment you can own. It lets you check connections, inspect speakers, and

Figure 2-26
DIN removal tools.
(*Courtesy of Scosche.*)

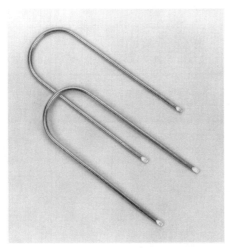

Recommended Test CDs

"My Disc"

From Sheffield Lab/Autosound 2000. Contains 86 tracks engineered to provide a thorough evaluation of your audio system's performance, the listening environment, and your personal critical listening ability. Contains special tracks that enable you to analyze performance with or without specialized test equipment. There are also six complete music tracks from Sheffield. A detailed 20-page booklet takes you through each procedure track by track.

The Sheffield Lab/Coustic Test Disc

Developed to help installers, competitors, and audio enthusiasts set up and test high-performance auto sound systems as well as home theater systems. Appropriate for the simplest of systems and for the most sophisticated. In addition to narration and musical excerpts, this disc provides test tones for proper channel identification, phasing, and polarity.

IASCA Set-up and Test CD

Recorded by Sheffield Lab and produced by Autosound 2000, this CD is the official International Auto Sound Challenge Association (IASCA) setup and test disc. It features 99 tracks to evaluate spectral balance, staging, imaging, linearity and noise, distortion, and frequency response. This disc will simplify the process of setting up and tuning a car audio system for competitive judging. Most of the tracks consist of actual music selections from the official IASCA judging disc.

Tip: If your multimeter comes with pointed tip probes, you might want to replace both of them (or just the black one) with an alligator clip or mini-hook-type test clip. This solves the frustrating problem of only having two hands when you're trying to simultaneously make good connections with two probes and read the meter.

verify power to equipment. Without it, tracking down problems is like working in the dark.

A multimeter can be bought for as little as $10, a digital model for as little as $20. I highly recommend digital. If you never make mistakes, never have equipment problems, and always have full wiring documentation for the vehicles you work on, then you don't need one. Otherwise, it's a worth-

Figure 2-27
Test CDs. (*Courtesy of Autosound 2000.*)

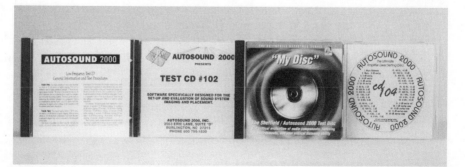

Figure 2-28
Digital multimeter.
(*Courtesy of MCM
Electronics.*)

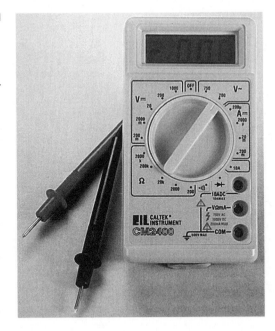

while investment. Buy a model with an audible continuity checker. Current measurement capabilities are not required.

Using a Multimeter—Wiring Harness Example

The basic functions of a multimeter are shown in Table 2-5.

The best way to explain how to use a multimeter is to dive right in with an example. The following shows how to use a multimeter to figure out a radio wiring harness in a car.

Let's say you plan to replace a factory radio with something better. You've removed the factory radio, and the eight-pin connector (which plugged into the back of the radio) looks like the one in Fig. 2-29.

Tip: For most vehicles, a wiring harness adapter is available. Wiring harness adapters mate with the existing car harness so you don't need to splice into any wires. Adapters are supplied with a color code cross-reference that identifies all the wires so you don't need to figure them out yourself.

Step 1: Identify the Power Wires

Turn on your meter and put it in the DC VOLTS mode. Connect the black probe of the meter to a known good electrical ground, such as bare

TABLE 2-5

Multimeter Functions and Their Uses

Function	Common Car Stereo Use
Measure DC voltage	Confirm that power is provided to a component
Measure AC voltage	—
Measure continuity	Verify a connection
Measure resistance	Check a speaker or speaker connection
Measure current	—

metal on the body of the car or the outer sleeve of the cigarette lighter socket.

Table 2-6 shows the three types of power wires that may be connected to a car radio.

Some or all of these may be provided to the radio. First, we'll look for the BATTERY +12V. With the headlights off and the key out of the ignition, probe each of the terminals inside the connector. In our example, we find that the red wire measures approximately +12 volts.

Next, turn on the headlights and repeat the procedure. In this case, we find that the yellow wire now measures about +12 volts. Finally, put the key in the ignition and turn it to the ACCESSORY or IGNITION position (you don't need to start the car). Now the orange wire measures +12 volts too, so it must be the accessory/ignition wire.

We have identified all the power wires (Fig. 2-30), so the lights can be turned off and the key taken out of the ignition.

Step 2: Identify the Ground Wire

Put your meter in the CONTINUITY mode. This mode is often indicated with a musical note symbol. If your meter doesn't have this feature, use the RESISTANCE mode instead. Momentarily connect the two

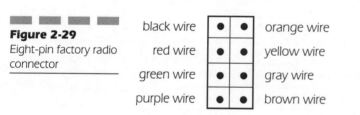

Figure 2-29

Eight-pin factory radio connector

TABLE 2-6

Types of Power
Wires in a Radio
Harness

Name	When On
Battery	Always on
Accessory/ignition	On when engine is running (or when ignition switch is in ACC position)
Lights	On when headlights (or dash lights) are on

probes of your meter together, and you should hear a beep. If you are using the RESISTANCE mode, you should see a meter reading very close to 0 ohms when the two probe tips are touching. If it doesn't read near zero or jumps around when you move the leads, you have a bad connection.

Once again, connect the black probe of the meter to a known good electrical ground, such as bare metal on the body of the car or the outer sleeve of the cigarette lighter socket. Now probe each of the wires in the connector that have not previously been identified. In our example, we find that the black wire causes the multimeter to beep, so it must be the ground wire (Fig. 2-31).

Step 3: Identify the Speaker Wires

By the process of elimination, the remaining wires must be speaker wires. Put the multimeter in the RESISTANCE mode.

Tip: Get in the habit of shorting your test probes together before making a resistance measurement to make sure that you get a meter reading near zero.

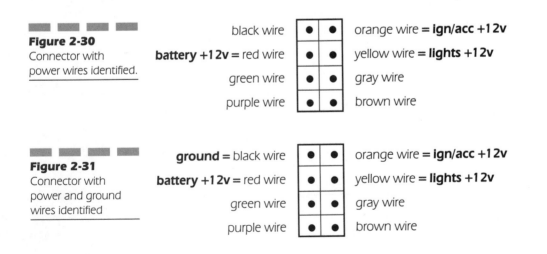

Figure 2-30
Connector with power wires identified.

Figure 2-31
Connector with power and ground wires identified

> **What Is Ground?**
>
> Modern cars use a 12-volt DC negative ground electrical system. (When the engine is running, this voltage can actually be as high as 14.4 volts.) This means that the frame, chassis, and body of the car, as well as almost all the electrical circuits in the car, are electrically connected together and to the minus terminal of the car battery. This super-connection is referred to as *electrical ground*.
>
> The advantage of having an electrical system with a common ground is a reduction in wiring. For example, a light bulb only needs one wire rather than two—the second terminal of the bulb is connected to the chassis of the car.

Tip: Most car speakers are 4 ohms, but some are 8 or even 10 ohms. In the case of a 4-ohm nominal speaker, 4 ohms is the impedance for music signals. A multimeter actually measures DC resistance, which is a lower number, such as 3.2 ohms.

It doesn't matter which wire we start with. Connect one probe to the green wire terminal. Now, use the other probe to search the other unidentified terminals for one that gives a meter reading between 3 and 10 ohms.

In this example, we find that the purple wire gives a meter reading of 3.5 ohms. (Some meters might show 0.0035 Kohms, which is equivalent, since 1 Kohm equals 1000 ohms.) The other terminals show infinite resistance with respect to the green wire. This reveals that the green and purple wires go to the same speaker.

The only unidentified wires remaining are the gray and the brown ones. Connecting a probe to each gives a meter reading of 3.5 ohms, confirming that these wires go to a second speaker.

We know that the green and purple wires go to one speaker and that the gray and brown wires go to another. We don't know the polarity of the wires or which wires go to which speaker. Sometimes you can tell by looking at the actual speaker connections inside the trunk or by removing a door speaker. At other times you may need to use the sophisticated speaker polarity tester in Fig. 2-32.

This is nothing more than a flashlight battery with wires connected. Any 1.5-volt battery will work. (Note: do not use a 9-volt battery to check

Figure 2-32
Speaker polarity tester.

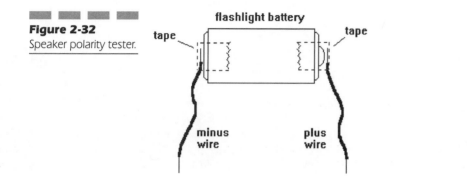

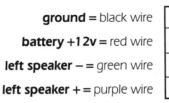

Figure 2-33
Connector with all wires identified.

ground = black wire	● ●	orange wire = **ign/acc +12v**
battery +12v = red wire	● ●	yellow wire = **lights +12v**
left speaker – = green wire	● ●	gray wire = **right speaker –**
left speaker + = purple wire	● ●	brown wire = **right speaker +**

a tweeter's polarity—the higher test voltage could destroy it.) Use tape to hold the wires onto the battery terminals. Connect the minus wire to the green terminal and touch the plus wire to the purple terminal. In this case, we hear a pop coming from the left speaker. Now watch the cone of the left speaker when you connect the battery. Does it move out or in? In this example, it moves out—this means that the plus terminal of the battery is connected to the plus terminal of the speaker. (If it moves in, the plus terminal of the battery is connected to the minus terminal of the speaker.) Repeating this procedure for the gray and brown wires, we complete our chart (Fig. 2-33).

CHAPTER **3**

Speakers and Speaker Projects

With today's factory stereos, speakers are often the weak link in system performance. The problem may be a lack of deep bass, a lack of a crisp high end, high distortion, or poor imaging. Adding or upgrading speakers can be the key to improving your system (Table 3-1).

In many cases, it makes sense to keep some or all of the existing speakers and correct their deficiencies by adding surface-mount tweeters or a subwoofer. In other cases, replacing the existing speakers with something better makes the most sense. For exceptional performance (and a price to match), separates can be used.

The following sections present various speaker projects independently. But, in many cases, information of general use may not be repeated in each section. For example, the section on upgrading existing speakers gives advice on choosing drivers. This advice applies just as well to other sections.

The imaging overhaul is presented first because it introduces imaging principles used in all later sections.

Imaging and You

What Is Imaging?

The term *imaging* refers to how well you can "close your eyes and tell where the sax player is standing." In your living room, proper imaging is easily achieved by arranging your speakers and listening chair as shown in Fig. 3-1.

In the recording studio, music is mixed to provide proper imaging for speakers arranged this way—in front of the listener, and forming a 60° angle. It makes sense to optimize music for this speaker arrangement because that's how most people listen to it. This arrangement also makes sense because it duplicates the listening zone of a normal live performance (assuming you have good seats).

Figure 3-1
Ideal listening room
arrangement.

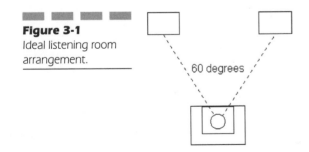

60 degrees

TABLE 3-1

Popular Speaker
Projects

Project	Comments
Imaging overhaul	Optimizes imaging—may require adding a subwoofer.
Adding surface mount tweeters	Low-cost way to improve high-frequency response of moderate-quality speakers.
Upgrading existing speakers	Simple and unobtrusive way to extend high-frequency and bass response plus increase power handling.
Installing speakers in new location	Useful when existing speaker locations are inadequate.
Installing separates	Provides highest sound quality, but expensive and labor intensive.
Adding a subwoofer	Best way to achieve thunderous bass—removes burden of bass from the other speakers, allowing them to play louder and with less distortion. (See Chap. 4.)

A typical listening arrangement for a car is shown in Fig. 3-2.

This setup has the potential to wreak havoc with imaging. For example, the sax player who was supposed to be standing directly in front of you might now seem to surround you. Or, if the back speakers are louder than the front, the sax player may seem to be directly behind you.

This leads up to the key question: How much do you care about imaging? For most people, imaging in the car is not a big concern. But for some, it is. Achieving good imaging is not without its sacrifices, either. It puts a big performance burden on the front speakers as primary drivers, since the rear speakers are used only for fill.

Figure 3-2
Typical car arrangement.

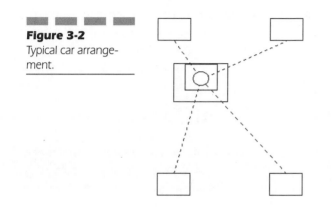

Achieving Good Imaging

For most vehicles, good imaging means sacrificing the easy bass available from trunk-mounted main drivers. For pickup trucks and mid-engine sports cars (where trunk-mounted speakers are not an option), this isn't as much of an issue. In either case, the difficulty of getting good bass out of the front usually makes a subwoofer necessary when good imaging is the goal.

For those who want to perform an imaging overhaul, here is my three-point program:

■ Use the front speakers as your primary drivers. The goal here is to provide good power handling and frequency response down to 150 Hz, where a subwoofer can take over. A crossover can be used to block energy below 150 Hz to the primary drivers. This allows them to play louder with less distortion. You may also decide to upgrade your existing front speakers with drop-in replacements or even go with separates.

■ Use the rear speakers for fill. Operating rear speakers at a reduced volume level (with respect to the front speakers) actually expands the stereo image. Rear speakers used for fill don't need to have extended bass, high frequency, or power handling capabilities, so don't buy expensive drivers for this application (you can keep the rear speakers you have). Set the front/rear fader control to the point where you can barely tell that the rear speakers exist.

■ Add a subwoofer. A subwoofer takes over the job of providing deep bass where the front speakers normally give out. Removing the burden of bass from the front speakers allows them to play louder and with less distortion. (Chapter 4 has details.)

In those rare cases where the front speakers provide adequate deep bass and power handling, you can consider yourself lucky and forget about adding a subwoofer.

Adding Surface-Mount Tweeters

Inexpensive speakers usually do a good job of reproducing mid-range, but they generally lack good high- and low-frequency capabilities. Adding surface-mount tweeters (Fig. 3-3) is a low-cost way to add a crisp

Figure 3-3
Surface-mount
tweeters. (*Courtesy of
JL Audio.*)

high end to cheap factory speakers not incorporating a tweeter. If the sound of cymbals doesn't shimmer, then your high end needs improvement.

Choosing Tweeters

A pair of surface-mount tweeter modules can cost anywhere from $12 to over $200. I have found that units costing as little as $20 a pair can provide excellent performance. Follow these rules to avoid buying something you'll regret later:

- Buy domes, not cones.
- Do not buy piezo tweeters.
- For high-power applications, ferrofluid is recommended.

Dome tweeters offer superior dispersion (which means they radiate well in all directions) and high-frequency response characteristics compared to cones. Cones are an obsolete technology for tweeters. Watch out—if it doesn't say it's a dome, it's probably a cone.

Piezo tweeters are highly efficient and do not require a crossover capacitor, but their advantages end there. Because of their inherently resonant behavior, they almost always exhibit a poor frequency response. Don't buy them.

Ferrofluid allows better cooling of the voice coil and thus provides greater power handling. It also smooths the frequency response somewhat. A ferrofluid-cooled model is recommended for high-power applications.

Another factor to consider is the mounting base. Some models allow the mounting angle to be adjusted by rotating the base. This can be quite helpful when trying to mount the tweeters so they aim toward the ears of the listener.

Placing Tweeters

If you use rear speakers only for fill, you can add surface-mount tweeters in the front only. If you listen to front and rear speakers at nearly equal volume levels, you should add tweeters in the rear too. This will improve the frequency response for back-seat listeners and will avoid the problem of "floating" stereo images in which instruments appear to move around depending on the frequency range they are producing. For optimum performance, tweeter modules should be mounted using the following guidelines:

- Mount tweeters near their corresponding full-range speakers.
- Do not mount where a passenger forms an obstacle between you and a tweeter.
- Do not mount too close to your ear (you'll hear only that speaker).
- Mount tweeters as close to ear level as possible.
- Aim tweeters toward listeners' ears.
- Beware of windshield reflections.

Some of these rules conflict in a normal installation, and the practical and aesthetic aspects of mounting must be considered as well, so you must weigh the relative importance of many factors.

Under no circumstances should a tweeter be mounted more than 6 inches from its corresponding full-range speaker. Violating this rule will degrade imaging. Restrict tweeter placement to a 6-inch range in all directions, then consider how to best satisfy the other guidelines. Possible locations include door (Fig. 3-4), dashboard, windshield pillar, and kick panel. Remember, you'll need to run wires to the tweeter from the full-range speaker, so consider ease of wiring.

Windshields can be thought of as sound mirrors for tweeters. If you imagine your tweeter as a lightbulb, then consider what you would see in the reflection of the windshield when sitting in the driver's seat. The apparent image of the tweeter is just as real and powerful as the tweeter itself. If you can see both the lightbulb and its reflection at the same time,

Figure 3-4
Door-mounted
tweeters. (*Courtesy of
Crutchfield.*)

you are in trouble—imaging will be smeared. The further apart the apparent image and the actual source, the worse the smearing. If you must mount a tweeter on the top of the dashboard, either mount it very close to the windshield or aim it so that either the direct path or reflected path will strongly dominate over the other.

Connecting Tweeters

Surface-mount tweeters should be wired as shown in Fig. 3-5.

The crossover capacitor allows only the high-frequency components of the signal to reach the tweeter. This prevents low frequencies from causing distortion in the tweeter or burning it out. The crossover frequency defines the dividing line between the high frequencies, which are passed to the tweeter, and the low frequencies, which are filtered out. The capacitor should be the non-polarized type (also called bipolar). Table 3-2 shows some common values of crossover frequencies and the corresponding crossover capacitor values for 4-ohm tweeters.

Figure 3-5
Wiring for surface-
mount tweeters.

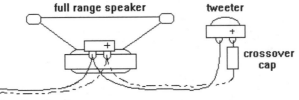

If you want to calculate your own capacitor values, use this equation:

$$C = \frac{1}{2\pi fR}$$

where f is the crossover frequency in hertz, R is the tweeter impedance in ohms, and C is farads.

Normally tweeter modules are provided with crossover capacitors, but you may wish to use a different value depending on the high-frequency performance of your existing speakers. Experiment to find the value that sounds best, opting for the smallest capacitor value (highest crossover frequency) you find acceptable.

After installing tweeter modules, you may find that they seem too loud compared to the existing speakers. (This is because tweeters are normally more efficient than full-range speakers.) There are three ways you can deal with this:

- Turn down the treble control on the head unit.
- Increase the crossover frequency (by reducing the capacitor value).
- Pad the tweeters using resistors.

Pads are special attenuation circuits designed to maintain the impedance looking into them. In this case, the pad maintains a 4-ohm input impedance so that the crossover works as intended. The circuit for padding tweeters is shown in Fig. 3-6. Table 3-3 shows the resistors to use for 3- and 6-dB pads for a 4-ohm tweeter. Be sure to use power resistors, or

Important: Changing the crossover frequency strongly affects how much power the tweeter must handle. For example, *lowering* the crossover frequency from 8 to 6 kHz subjects the tweeter to roughly 40 percent *more* power. Similarly, *increasing* the crossover frequency from 8 to 12 kHz subjects the tweeter to roughly 40 percent *less* power.

TABLE 3-2

Capacitor Values
for Common Cross-
over Frequencies

Crossover Frequency	Capacitor Value (4-ohm Tweeter)
6 kHz	6.8 µF
8 kHz	4.7 µF
12 kHz	3.3 µF

Figure 3-6
Circuit for padding
tweeters.

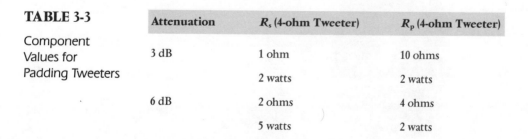

tweeter

full range speaker

Rp

Rs

crossover
cap

from head unit

you will burn them out. The power ratings shown are adequate for a
high-power head unit—power ratings will need to be increased when
higher-power amplifiers are used. Mount the crossover and pad components on a small board secured to the car. This will prevent short circuits
to ground that could blow your head unit.

Upgrading Existing Speakers

Upgrading your existing speakers is a simple and unobtrusive way to
extend high-frequency and bass response. It also lets you increase power
handling.

Choosing Speakers

The first step in upgrading is to understand what you have space for.
Drop-in replacements are made for almost any size opening, but you
must know the maximum depth as well as the size of the mounting hole.
For odd-size factory holes, mounting adapter plates that permit standard
aftermarket speakers to fit are available.

TABLE 3-3

Component
Values for
Padding Tweeters

Attenuation	R_s (4-ohm Tweeter)	R_p (4-ohm Tweeter)
3 dB	1 ohm	10 ohms
	2 watts	2 watts
6 dB	2 ohms	4 ohms
	5 watts	2 watts

Figure 3-7
Kappa 6912i two-way
speaker. (*Courtesy of
Infinity.*)

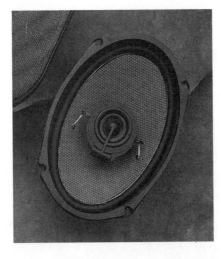

Crutchfield has done a tremendous job in this department. Their cata-log shows what size speakers (including depth) fit in each factory location in your vehicle. They also provide installation guides that tell you how to access and remove your existing speakers. Their product lineup includes a wide selection of speakers.

When choosing speakers, opt for a two-way model using a dome tweeter (Fig. 3-7). Dual-cone models can't reproduce high frequencies, and the extra driver in three-way models (Fig. 3-8) is more for show than per-formance. Power handling and frequency response are important, but unfortunately you can't trust most published specs. Even supposedly rep-utable manufacturers often make exorbitant claims about power handling and frequency response. All other factors being equal, voice coil diameter is a good indicator of power handling. A typical 1-inch voice coil can be expected to handle about 35 watts, and a 2-inch voice coil about 125 watts.

Figure 3-8
Kappa 6931i three-way speaker. (*Courtesy of Infinity.*)

TABLE 3-4

Common Speaker
Cone Materials

⟨BEST⟩

Cone Material	Comments
Paper	Deteriorates with sunlight and moisture
Coated paper	Good choice
Polypropylene	Temperature-sensitive performance

For frequency response, there's something to be said for listening to speakers in a showroom.

Sensitivity is a term for how loud a speaker will play with 1 watt of power. This is important if you are using a normal or high-powered head unit, but not so important if you are using separate power amps capable of higher output power. A decibel or two of difference doesn't matter, but more than that should be weighed in your purchase decision.

Cone material (Table 3-4) and surround material (Table 3-5) are important in the harsh automotive environment, but it's hard to wade through the hype. Polypropylene cones are extremely resistant to environmental deterioration, but are a temperature-sensitive plastic, becoming soft in the heat and hard when cold. Paper cones deteriorate with sunlight and moisture, so you must keep them out of the sun. Coated paper is a good compromise. Other materials such as graphite, Tri-Laminate, resin laminate, carbon-blended poly, kapok, poly-graphite, graphite-quartz, Foam-Infused IMPP, titanium composite, fiberglass, and Kevlar offer the promise of superior performance. Don't expect a big improvement over coated paper.

The surround is the soft ring around the outside of the cone with the bulge in it. Choose rubber if this driver will be exposed to the sun (for example, if you're mounting it on the top of the dashboard or on the rear deck).

Strive for the most bass possible in front. It will prevent the undesirable effect of all the bass coming from behind you. The keys to good bass in the front (or anywhere, for that matter) are a good driver and a good enclosure. Creating good bass means moving lots of air, so cone area is important too. Choose the largest driver that will fit in a hole. You can buy adapters that let you mount a larger driver in a hole (Fig. 3-9): for example,

TABLE 3-5

Common Speaker
Surround Materials

⟨BEST⟩

Surround Material	Comments
Foam	Deteriorates with sunlight
Butyl rubber	Good choice

Figure 3-9
6 × 9 adapters.
(*Courtesy of MCM Electronics.*)

Figure 3-9
6 × 9 adapters.
(*Courtesy of MCM Electronics.*)

a 6 × 9 driver in a 4 × 10 hole. Depth extender rings that let you install deep speakers in shallow compartments such as kick panels or doors are also available (Fig. 3-10).

Making the Best of the Existing Enclosure

The enclosure will strongly affect your sound quality, especially the bass and mid-range. To a large extent, you are stuck with whatever the car has to offer. Ideally, the enclosure is large, sealed, and nonresonant.

In the case of replacing door speakers, doors are usually large and somewhat sealed, but they are typically resonant. You can reduce the resonance problem by using foam baffles (Fig. 3-11). These are soft foam cups installed behind the drivers. Foam baffles have the added benefit of reducing road noise (which can pass right through the speaker cone) as well as protecting the back of your speaker against water and dirt that may sometimes find their way inside your door.

In the case of replacing dash-mounted speakers, the enclosure is an open-backed cabinet. Open-backed cabinets allow the sound from the back of the speaker to be heard—in this case, from under the dashboard. This causes frequency response ripples (peaks and valleys) in the mid-range and cancellation of deep bass. The closer the speaker is to the bottom of the dashboard, the worse the problem is.

Figure 3-10
Depth extender rings.
(*Courtesy of MCM Electronics.*)

Figure 3-11
Foam baffles. (*Courtesy of MCM Electronics.*)

Some improvement can be made in the case of frequency response ripples by installing a cardboard box stuffed with damping material (such as fiberglass or polyester batting) behind the driver (see Fig. 3-12). You may be able to use the flaps of the box as a mounting flange; otherwise you must devise your own method of securing the box. It's not important that the box make a tight seal against the dashboard, but it should be secure enough not to rattle or vibrate.

In cases where dash speakers are mounted near the bottom of the dash, it may be best to consider an alternate location.

Installing Speakers in a New Location

Installing speakers in a new location is necessary when existing speaker locations are inadequate. Existing locations may be inadequate for reasons of imaging, inability to accommodate drivers of the desired size, or enclosure volume. This is often the case for front factory speaker openings.

One important rule when adding speakers is that more isn't better. This refers to the number of speakers covering a particular frequency range.

Figure 3-12
Box with damping material behind the driver.

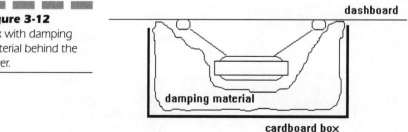

For example, you don't want to have a full-range driver in the kick panel and also on the top of the dashboard. Even with good-quality drivers, you should not try to create a wall of sound. Not only will it degrade your imaging, it will create a frequency response with lots of peaks and valleys. (This is because the two drivers have a slightly different time delay to reach your ear, and this produces an interference pattern that cancels some frequencies and reinforces others.) If you need increased power handling, choose a speaker that can provide it rather than using multiple drivers.

The one exception to the "more isn't better" rule is subwoofers. The wavelengths at deep bass frequencies are long compared to the distances between drivers, so interference patterns aren't a problem.

Choosing a Location

There are two main factors for deciding where to put speakers for the best sound: imaging and enclosure. Imaging depends on where the speakers are with respect to your ears. It also involves things like reflections off the windshield or obstructions, such as passengers. Enclosure is the space that the back of the speaker radiates into—ideally it's large, sealed, and non-resonant. Table 3-6 shows common mounting locations for front speakers and their imaging and enclosure traits.

The front edge of the front doors is probably the best overall location for front speakers. It provides both a good enclosure and good imaging location. Placing the speakers toward the center or rear of the front doors

TABLE 3-6

Common Front
Speaker Locations

Front Speaker Location	Comments
Front door—front	Good location
Front door—center or rear	Good enclosure but very poor imaging location
Kick panel	Small enclosure makes bass below 150 Hz difficult: For good imaging, must aim toward ears
Dashboard—top	Windshield reflections degrade imaging: Open-back enclosure produces mid-range frequency response ripples and makes bass below 100 Hz difficult
Dashboard—front or bottom	Open-back enclosure produces strong mid-range frequency response ripples and makes bass below 200 Hz difficult

BEST

causes each listener to hear mainly one speaker and shifts the image too far back. In some cases, it may cause a passenger to block a speaker entirely. Use foam baffles behind door-mounted drivers to reduce resonance problems and road noise as well as to protect the back of your speakers against water and dirt.

Kick panels (the side panels under the dash) can be acceptable mounting locations, but speakers here should be aimed toward the listener for good imaging. The small enclosure volume usually strongly limits the bass response. This means you will either need to use back speakers to provide deep bass or add a subwoofer.

The dashboard is, in general, not a preferred mounting location because it is an open-backed cabinet. This allows the sound from the back of the speaker to be heard through the bottom of the dashboard. The combined sound from the front of the speaker and under the dash results in a frequency response with ripples in the mid-range and deep bass cancellation. The cardboard box with damping material, explained previously, helps with mid-range smoothness. Because of the limited bass response, you will either need to use back speakers to provide deep bass or add a subwoofer.

The best location for rear speakers should be determined by enclosure rather than imaging concerns. This is because in most installations the rear speakers provide the deep bass for the entire system. You should choose whatever location provides the biggest and best sealed box. If you have a trunk, use it!

Installing a Door Speaker

The first step in installing a new speaker in a door is removing the door panel. Start by removing the window crank handle (Fig. 3-13). Some handles are held on with a screw, which is sometimes hidden by a snap-on

Figure 3-13
Removing a window crank. (*Courtesy of Crutchfield.*)

cap. Most are held in place by a spring clip. To remove the clip, you can use a window crank clip removing tool or a piece of coat hanger with a tiny hook bent into the end. Depress the surrounding panel, look behind the window crank handle, and rotate the handle until you see the spring clip. Then gently push it off with the window crank clip removing tool or pull it off with the hook.

NOTE   *To reinstall the handle later, first snap the spring clip onto the handle, then push the handle back onto its shaft until it locks into place.*

After removing the window crank handle, remove the armrest (usually secured with a few Phillips-head screws) and any trim around the door handle. The only thing holding the door panel on now should be a half dozen friction fittings and possibly a few more screws.

With all the screws removed, start at a bottom corner and pull the panel straight out. Use a large flathead screwdriver to help pry it off. If a friction fitting breaks, you can get replacements from an auto parts store. Once the corner is loose, work across the bottom and up both sides, being careful not to use too much force. With the bottom and sides loose, the panel should now be hanging by some trim that sticks down into the window well. Lift straight up and it should come free.

Cutting the Hole

Once you've decided on the general location for the new speakers, get out the template that comes with them. You'll find it along with the instructions or printed on the box. Place the template over the area you've selected and make note of any possible obstructions or problems such as an irregular mounting surface.

Check what's behind the surface you're about to cut. Make sure no internal mechanisms will be affected. Be especially careful to check that the speaker will not interfere with the window and window crank mechanisms. Check the clearance with the window rolled down. If it's close, try rolling the window up and down before doing anything permanent. Also try closing the door—the mechanism that holds the door open can sometimes intrude into the speaker mounting cavity as the door closes. The risk of clearance problems is smaller when enlarging a factory location for a speaker than when creating a new opening.

Keeping in mind the general area you've selected for the speaker, check the door to see how much, if any, metal you'll need to cut. Very few doors are solid sheets of metal—there are usually several holes. Double-check the depth available in the door for mounting your speaker. If you think the measurement is too close, you may be able to use a spacer ring.

Locate this same position on the back of the door panel. Tape the supplied template on the exact spot and trace the inside edge with a pen. Lay the panel flat on a clean surface and cut out the circle with a sharp X-Acto or utility knife. Be patient. A dull blade or too much haste here might rip the fabric. Repeat the procedure for the other door, making sure to cut in the same place. You can hold the cut door panel against the uncut one and trace the hole with a pen to guarantee they will be the same.

Metal Cutting Tips

Here are some techniques that will help you make cleaner cuts and avoid damaging your car:

- Instead of trying to cut through the door's hardboard panel and sheet metal at the same time, remove and cut the hardboard panel with an X-Acto or utility knife. Then cut the corresponding sheet metal with a jigsaw equipped with a metal cutting blade. If you try to cut both layers at once, you may rip the panel covering.
- Make sure the blade of the jigsaw (in its most extended position) doesn't reach through the surface you're cutting to the exterior sheet metal of the car. If it does, it will peck numerous dents very quickly! Similarly, make sure the window is rolled up and there are no other obstructions the saw blade can reach.
- If you're cutting metal that will not be concealed behind a door panel, wrap the base of the saw with electrical tape to prevent marring.
- When using a jigsaw or any other power tool, always wear safety glasses or goggles (Fig. 3-14).

Figure 3-14
Cutting a door hole with a jigsaw. (*Courtesy of Crutchfield.*)

Frequently the speaker location will partially overlap an existing hole, which can be easily enlarged. If the amount of cutting involved is small, a utility hacksaw can be used. If there is no hole to start with, use a drill to make a pilot hole for the jigsaw blade.

After the hole is cut, but before the speaker is installed, vacuum out all the metal particles and other debris. Particles can get into a speaker and cause buzzing.

Wiring It Up

Before replacing the door panel, don't forget to run the speaker wire. It will have to exit the body of the car in the doorjamb and enter the door as close to that spot as possible. Often there are precut holes (with plugs in them) that will work out nicely. To run your wires from the door into the car body, try to use the factory rubber tubing between the door and the door jamb. If the tube is not present on your car, it may be available from a junkyard. If a factory boot or plugs are not present, you'll need to drill ⅜-inch holes to run the wiring. Before you drill a hole in the door, make sure it will provide access to the speaker location. Sometimes structural steel isolates the front edge of the door. It may still be possible to route this way, but check first.

Protect the wiring from the sharp edges of the holes by using rubber grommets in the holes or flexible tubing run between the two holes. This will keep the insulation from being cut after countless openings and closings of the door. Make sure the wire can't get pinched by the hinge or some other portion of the door jamb. Remember to leave enough slack in the wire to accommodate the door opening all the way.

Once you've finished pulling the speaker wire through the holes, pull the end through the speaker hole you've cut in the door panel and hang the panel from the top (hooked into the window well). If you accidentally broke a strategically located (like a corner) friction fitting when removing the panel, replace it with either a new one, or in a pinch, one taken from the middle of the panel along the bottom.

If your new speaker is designed to be top mounted, you can replace the door panel now. A foam baffle should be installed between the door and the door panel. When the door panel is properly positioned, push the friction fittings back into their seats and replace enough of the screws, armrest, etc. to hold the panel loosely in place. Do not reattach the window crank yet, since you may have to remove the panel again.

Hold the speaker in its new home, mark the screw holes, and remove the speaker. Drill the holes. Connect the speaker wires to the speaker, observing the proper polarity. The positive terminal is usually marked with a + or a colored dot.

Speaker Polarity Tip

It is important to observe the correct speaker polarity when connecting speaker wires. If you attach a set of wires to the terminals of one speaker backward, that speaker will be out of phase and your bass performance will suffer as a result.

NOTE If EVERY speaker in the vehicle is wired backward, no problems will result, but it is best to wire them properly to make sure future repairs or speaker additions won't inadvertently cause a phasing problem.

If the positive terminal of a speaker isn't marked, use the speaker polarity tester in Fig. 3-15.

This is nothing more than a flashlight battery with wires connected. Any 1.5-volt battery will work. (Note: Do not use a 9-volt battery to check a tweeter's polarity—it could destroy the tweeter.) Use tape to hold the wires onto the battery terminals.

Connect the minus wire to one speaker terminal and touch the plus wire to the other speaker terminal. Watch the speaker cone when you connect the battery. Does it move out or in? If it moves out, then the plus terminal of the battery is connected to the plus terminal of the speaker. (Otherwise, the plus terminal of the battery is connected to the minus terminal of the speaker.) Mark the plus terminal accordingly.

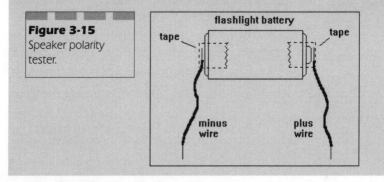

Figure 3-15
Speaker polarity tester.

When supplied, use speed clips (Fig. 3-16) over the new screw holes. They give the screws something extra to hold on to, providing extra support when the door is slammed. Once the speaker is installed, put the grille on immediately to prevent damage to the speaker while you do other work.

Figure 3-16
Speed clip.

Installing Separates

Separates (also called component speakers) are individually installed woofers and tweeters (Fig. 3-17). (This is instead of having the tweeter mounted in front of the woofer, as is the case with coaxial speakers). Normally, separates are sold as matched sets, including woofer, tweeter, and crossover.

The advantages of separates are the quality of the components—especially the crossover—and the flexibility of mounting. The disadvantages are the high cost and the additional installation work.

The quality of component woofers and tweeters is usually superior to that of coaxial speakers, for marketing rather than technical reasons. In other words, there's no reason why component drivers should inherently be any better than those used in a coaxial speaker, other than that this market niche demands it. So don't rush out and buy some amazingly low-priced component system assuming you're getting more than you paid for.

The crossover splits the music signal into low frequencies (bass), which go to the woofer, and high frequencies (treble), which go to the tweeter. In a typical coaxial speaker, a simple capacitor is used to prevent bass from going to the tweeter and nothing is used to prevent high frequencies from going to the woofer. This approach is inexpensive and small enough to fit conve-

Figure 3-17
Separates. (*Courtesy of JL Audio.*)

niently on the basket assembly of the speaker. Unfortunately its performance is marginal. The crossover module provided with component systems does not have the same size limitation, and because of the marketing niche, some money can be spent on a real crossover. Component crossovers provide filtering for both the woofer and tweeter and are often 12 dB/octave (as opposed to 6 dB/octave for a simple capacitor). The difference can be quite audible, particularly in mid-range smoothness and lower distortion.

Having a separate tweeter provides flexibility of mounting in the sense that you can now mount the woofer in a spot that would be unacceptable for tweeter orientation in a coaxial speaker. Also, you can mount the tweeter in a spot where a woofer wouldn't fit. For example, you could mount a woofer on the bottom front corner of the kick panel, facing the opposite woofer, and a tweeter on the top rear corner of the kick panel, facing the listeners.

Follow the guidelines in the "Adding Surface-Mount Tweeters" section earlier in this chapter for where to mount tweeters. Remember, a tweeter should not be mounted more than 6 inches from its woofer. Violating this rule will degrade imaging. Restrict tweeter placement to a 6-inch range in all directions, then consider how to best satisfy the other mounting guidelines.

Separates are most advantageous in front, where mid-range performance is most critical and mounting flexibility is more likely to provide a benefit. You may choose to use separates in front and coaxial speakers in the rear to save money without sacrificing top-notch performance.

If you plan to use separates in the front, consider the kick panel pods by Q-Forms (Fig. 3-18). These custom-molded pods are available for many

Figure 3-18
Q-Forms kick panel pods—(*a*) before; (*b*) after. (*Courtesy of Ai Research.*)

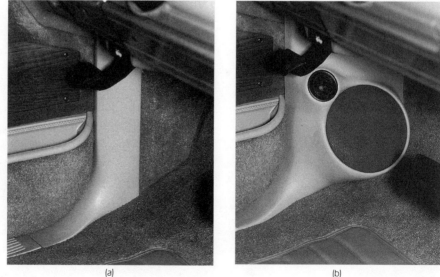

(a)   (b)

popular vehicles and accommodate a 5¼-inch woofer plus a separate tweeter. The pods aim the speakers toward the listener for good imaging, increase the size of the enclosure to improve bass response, and maintain a factory appearance. They are available from Crutchfield and are priced at about $200 a pair.

CHAPTER 4

Subwoofers and Subwoofer Projects

This chapter is by far the largest in the book. There are two good reasons for this. First, subwoofers ("subs") are one of the most important components in a high-quality auto sound system. Second, there are a tremendous number of options you must choose from every step of the way.

A subwoofer not only gives you a rock-solid bottom end, it lets you play your other speakers louder with less distortion by diverting the burden of deep bass to the subwoofer. A subwoofer is almost a necessity if you use the front speakers for imaging and the rear speakers for fill, because of the difficulty in getting decent bass from the front.

Because subs are so important, manufacturers offer a staggering array of products to support a vast number of approaches. You must choose between everything from powered subs with built-in crossovers to separates, and every combination in between. You can design and build your own box, choose a pre-made one and install your own driver, or buy a ready-made system. There are sealed, ported, bandpass, and other types of enclosures to choose from.

This chapter will cut through the hype and give you the know-how to choose and use the right components, whether you want to buy or build.

System Configuration Options

There are six basic configurations for adding a subwoofer system (Table 4-1). The first two use *speaker-level* crossovers, the rest use *preamp-level* crossovers.

Speaker-level crossovers are used between the amplifiers and speakers and are passive (requiring no power). Preamp-level crossovers are used before the amplifiers and are generally active (requiring power). It's common to see the terms *passive* and *speaker-level* used synonymously in advertising, but preamp-level crossovers can be passive too.

The preamp-level crossover configurations vary only in how they repackage the three basic building blocks of a subwoofer system: crossover, amp, and subwoofer. Separates provide the greatest flexibility; amplified subs with built-in crossovers provide the most convenience. Other options fall in between.

The secret to good subwoofer performance is in the crossover. Regardless of which approach you choose, make sure you satisfy the requirements for a good crossover. This is explained in detail in the section entitled "What to Look For in a Subwoofer Crossover," later in this chapter.

TABLE 4-1

Configurations
for Adding a
Subwoofer

Configuration	Diagram	Comments
Conventional speaker-level crossover, sub		Lets you use a single amp to drive a dual voice coil subwoofer and boost a pair of speakers. Subwoofer level and crossover frequency are not adjustable. Typical 12-dB/octave sub slope is inadequate for most applications. Not recommended.
Tri-way crossover, sub		Lets you use a single tri-way-capable amp to drive a sub-woofer and boost a pair of speakers. Subwoofer level and crossover frequency are not adjustable. Typical 12-dB/octave sub slope is inadequate for most applications. Not recommended.
Separates: crossover, amp, sub		Offers the most flexibility and best performance. (Amp has no built-in crossover or one considered inadequate.)
Amp with built-in crossover, sub		Can save the cost of a separate subwoofer crossover. Built-in crossover may be inadequate.
Separate crossover, amplified sub		Offers potential cost savings and convenience by combining a subwoofer with a matching amp. (Amplified subwoofer has no built-in crossover or one considered inadequate.)
Amplified sub with built-in crossover		The ultimate in convenience. Built-in crossover may be inadequate.

Conventional Speaker-Level Crossover, Subwoofer

A conventional speaker-level crossover connects between any power amp and your speakers. It's a high-power device that generally provides a 6-dB/octave high-pass filter to a pair of main speakers and a 12-dB/octave low-pass filter to a dual voice coil subwoofer (or a pair of regular sub-woofers).

This approach lets you use a single amp (rather than two) to both drive a subwoofer and boost a pair of speakers. This would seem to provide a substantial cost savings, but for the same amount of total power, this is not usually the case. For example, the cost of a single $100W \times 2$ amp for a speaker-level crossover setup is comparable to the cost of two $50W \times 2$ amps for a preamp-level crossover approach.

In addition, there are a number of performance limitations involved with using a speaker-level crossover. Because a single amp is shared by all drivers, the tweeters are unprotected against clipping of deep bass to the subwoofer (explained in detail in Chap. 8). Unlike most preamp-level crossovers, speaker-level crossovers have fixed crossover frequencies and cannot be adjusted to optimize your system. Furthermore, the subwoofer level is not adjustable with respect to the main speakers with a speaker-level crossover.

Finally, the 12-dB/octave subwoofer slope usually found in this type of crossover is generally inadequate. This is because a 12-dB/octave slope allows mid-bass to be heard, which lets you hear where the sub is located. This degrades imaging. You can fix this deficiency by using an 18- or 24-dB/octave slope, but this generally requires you to design and build your own crossover using coils and capacitors. A 12-dB/octave slope can be acceptable with bandpass subwoofers because of their inherently reduced high-frequency output.

For these reasons, conventional speaker-level crossovers are not recommended for subwoofers.

Tri-Way Crossover, Subwoofer

A tri-way (or tri-mode) crossover connects between a tri-way-capable amp and your speakers. It's a high-power, passive device that generally provides a 6-dB/octave high-pass filter to a pair of main speakers and a 12-dB/octave low-pass filter to a single subwoofer.

This approach lets you use a single amp (rather than two) to both drive a subwoofer and boost a pair of speakers. This would seem to provide a substantial cost savings, but for the same amount of total power, this is not usually the case. For example, the cost of a single 100W × 2 amp for a tri-way setup is comparable to the cost of two 50W × 2 amps for a preamp-level crossover approach.

In addition, there are a number of performance limitations involved with using a tri-way crossover. Because a single amp is shared by all drivers, the tweeters are unprotected against clipping of deep bass to the subwoofer (explained in detail in Chap. 8). Unlike most preamp-level crossovers, tri-way crossovers have fixed crossover frequencies and cannot be adjusted to optimize your system. The subwoofer level is not adjustable with respect to the main speakers in a tri-way setup. Finally, the 12-dB/octave subwoofer slope found in this type of crossover is generally inadequate. This is because a 12-dB/octave slope allows mid-bass to be heard, which lets you hear where the sub is located. This degrades imaging.

For these reasons, tri-way crossovers are not recommended.

Separates

Using a separate crossover, amp, and subwoofer offers you the most flexibility and the best performance of any of the configurations. You can choose whatever power level you need in an amplifier and pick a subwoofer that is a perfect fit for your vehicle. Most importantly, you can select a subwoofer crossover that provides a steep slope as well as important features usually found only in separate crossovers.

The amplifier in this configuration either has no built-in crossover or has one that is considered inadequate. You may be able to take advantage of an inadequate built-in crossover to reduce the requirements for a separate crossover. See the box to learn how.

Using a Built-In Crossover Together with a Separate Crossover

If you've got an inadequate built-in crossover, you have two choices:

- Bypass the built-in crossover and use a good separate crossover.
- Use the built-in crossover together with a separate crossover.

An inadequate built-in subwoofer crossover can be defeated by choosing the bypass setting. (If there is no bypass setting, setting the built-in crossover to the highest cutoff frequency will usually allow the separate crossover to dominate.) Since the separate crossover will now be doing all the work, you will need to choose a model that meets all your needs. The advantage of this approach is that you then have a single set of controls to deal with, and what you see with them is what you get.

Alternately, you can use the built-in crossover to effectively increase the performance of the separate crossover. This reduces the requirements for the separate crossover and can save money. For example, each crossover by itself might have a 12-dB/octave slope. Using the two together, you can achieve a 24-dB/octave slope (the slopes of two cascaded filters add).

Table 4-2 shows an example of how this works. Suppose both your crossovers are 12 dB/octave and have 80/120-Hz cutoff frequency selection switches. The possible crossover setting combinations and their combined responses are shown in Table 4-2.

TABLE 4-2

Combined Cutoff Frequency of Cascaded Crossovers

Separate Crossover Setting (12 dB/octave)	Built-In Crossover Setting (12 dB/octave)	Combined Filter Response (24 dB/octave)
80 Hz	80 Hz	65 Hz
80 Hz	120 Hz	74 Hz
120 Hz	80 Hz	74 Hz
120 Hz	120 Hz	97 Hz

There are two important things to notice in this table. First, you now have more cutoff frequencies available than either individual crossover provides. This gives you finer control (unless one of the crossovers was already continuously variable).

Second, the crossover frequencies shift to lower values when you combine two crossovers—about 20 percent below the nominally selected frequencies in this case. (This makes sense if you think about it: The two 80-Hz filters were each 3 dB down at 80 Hz, so the combination is 6 dB down at 80 Hz. The 3-dB down frequency of the combined response must be a lower frequency.) This could be good news or bad news depending on the frequency you want and the available frequencies.

Important: For best results with this approach, try to set both cutoff frequencies the same. This will give you the sharpest filter characteristic near the cutoff frequency.

Filter slopes aren't the only things that can benefit from this tactic. Features such as subsonic filters, polarity switches, and 45-Hz bass boost might be included in one crossover, but not the other.

Amp with Built-In Crossover, Subwoofer

Many of today's amps include a built-in low-pass crossover. This can be a great way to save the cost of a separate subwoofer crossover, but you need to make sure the built-in crossover meets your needs. (If it doesn't, consider the previous configuration and see the preceding box.)

Many built-in crossovers use shallow 6- or 12-dB/octave slopes. This may be acceptable if you plan to use a bandpass subwoofer (because of their inherently reduced high-frequency output), but shoot for 18 or 24 dB/octave otherwise. Similarly, many built-in crossovers offer only a single fixed cutoff frequency. This limits your ability to seamlessly integrate a subwoofer into your system. Few, if any, models offer subsonic filtering.

Amplifiers with capable built-in crossovers do exist. Some Kicker® amp models feature plug-in crossover modules providing 24-dB/octave slopes and selectable frequencies. Equalization modules are also available to add deep bass kick.

Separate Crossover, Amplified Subwoofer

This approach uses a separate crossover in conjunction with an amplified subwoofer. The amplified subwoofer either has no built-in crossover or has one that is considered inadequate. Either way, this approach offers potential cost savings and convenience by combining a subwoofer with a matching amp. Using a separate crossover guarantees you can obtain the performance and features you need in a crossover without severely limiting your choice of amplified subs.

Amplified subs without built-in subwoofer crossovers are becoming rare as manufacturers realize that they can include a bare-bones crossover for a small incremental cost. Unfortunately, most of these crossovers are inadequate for the job and need the help of an external crossover. You may be able to take advantage of an inadequate built-in crossover to reduce the requirements for a separate crossover. See the box on p. 59 to learn how.

The separate crossover used with this approach may be a subwoofer-only crossover or one with high-pass filtered outputs for the main channels too. If you are using separate amps to drive the main channels, then the latter is the way to go.

Amplified Subwoofer with Built-In Crossover

This approach is the ultimate in convenience. You just connect power and signal leads, adjust your settings, and you're done.

The main concern with this configuration is making sure the built-in crossover meets your needs. (If it doesn't, consider the previous configuration and see the box on p. 59.) Many of the built-in crossovers use shallow 6- or 12-dB/octave slopes. This may be acceptable for bandpass subwoofers, but you need 18 or 24 dB/octave otherwise.

Adding a Subwoofer to Premium Factory Sound Systems

If you have a premium factory sound system (such as Delco/Bose or Ford/JBL) and want to add a subwoofer, the situation is slightly more complicated than with standard systems.

You can use any of the four preamp-level crossover configurations outlined at the beginning of this chapter, but you'll need to use a line output converter designed specifically for Bose or Ford Premium systems. The line output converter is needed to convert the signals from the premium factory head unit to a standard preamp-level signal suitable for crossover inputs.

Line Output Converters for Bose/Ford Premium Head Units

Line output converters are commonly used to convert head unit speaker-level outputs to levels suitable for the preamp-level inputs of amps.

If you're dealing with a Bose or Ford Premium Sound factory head unit, you'll want to use a line output converter specifically designed for interfacing to these units. Using a standard LOC having low input impedance may damage the output circuit of a Bose or Ford Premium Sound head unit. Using a standard LOC with a Bose or Ford system can also result in extremely low output from the LOC because of the relatively low signal levels of many premium factory system head units.

The LOCB by Soundgate will accept any input signal from 175 mV to 6.6 volts and convert it to a 2.5-volt audio output suitable for aftermarket crossovers and amps. It also has a noise-blanking circuit that activates the audio path after turn-on noises have subsided.

Chrysler/Infinity Systems

Unlike Bose and Ford, Chrysler/Infinity systems use standard head unit technology. No line output converter is needed with crossovers having speaker-level inputs. A standard LOC will do the job of interfacing a Chrysler/Infinity head unit to a crossover lacking speaker-level inputs.

What to Look For in a Subwoofer Crossover

There are a number of considerations for choosing the right crossover for your application. You might choose a dedicated subwoofer crossover with subwoofer outputs only, or one with high-pass outputs too if you use amps to drive your main speakers. If you're thinking of biamping later on, you might choose a three-way model with subwoofer, woofer, and tweeter outputs. (See Chap. 8 for more details.)

The most important subwoofer crossover characteristics are listed in Table 4-3.

The crossover filter slope may be the single most important item to consider. To make it possible to put a subwoofer anywhere in the vehicle, the subwoofer crossover must be 18 dB/octave or higher. Slopes less than 18 dB/octave allow mid-bass and even mid-range to be heard from the subwoofer. Since only deep bass frequencies are nondirectional, this gives

TABLE 4-3

What to Look For in a Subwoofer Crossover

Item	Comments
Crossover filter slope	18-dB/octave minimum. 12-dB/octave minimum for bandpass boxes.
Cutoff frequency range	75- to 150-Hz minimum range. Continuously variable—a big plus.
High-pass outputs	Important if you have amps driving main speakers. 6-dB/octave slope minimum.
Subsonic filter	Crucial for ported boxes, beneficial for others.
45-Hz bass boost	Useful for improving bass.
Polarity switch	Useful for improving bass.
Speaker-level inputs	Important if your head unit lacks preamp-level outputs.
Preamp-level inputs	Ground loop isolation is a big plus for reducing noise problems.
Output level controls	Of limited value—usually redundant with amp input level controls.
Automatic amp wake-up output	Saves having to run remote turn-on wire from head unit to sub amp.

away the subwoofer location and degrades imaging. Who wants a bass guitar player under the seat? The only exception to this is bandpass boxes, where you can successfully use a 12-dB/octave crossover.

A subwoofer cutoff frequency range of 75 to 150 Hz in three or four steps is adequate for most situations. A continuously variable cutoff frequency control gives you full flexibility to seamlessly blend the subwoofer with the rest of your system. Even with high-pass crossovers on your main speakers, the natural low-frequency rolloff of woofers makes it difficult to predict the best subwoofer crossover frequency. A continuously variable crossover lets you avoid a gap or peak in your mid-bass.

High-pass outputs are an important feature if you have amps driving your main speakers. Using this feature saves you from having to install bass blocking crossovers in line with your main speakers. A 6-dB/octave slope is acceptable, but 12 dB/octave or higher is better. The high-pass slope for the main speakers does not need to be the same as the subwoofer low-pass slope.

A subsonic filter prevents very low frequencies from getting to the subwoofer amp and speaker. These frequencies are too low to be effectively reproduced by a subwoofer anyway, but they use up valuable amplifier and speaker headroom. A subsonic filter protects against unnecessary distortion and speaker damage as well as reducing annoying turn-on thumps. A subsonic filter is important for any subwoofer system, but it is crucial when ported boxes are used because they lack the protective air "shock absorber" of a sealed box.

Five to 10 dB of bass boost at 40 or 45 Hz extends the low-frequency response of most subwoofer systems without overdriving them.

A subwoofer polarity switch lets you select the proper polarity without having to reverse any speaker wires. Choose the setting that gives you the most bass.

If your head unit lacks preamp-level outputs, then speaker-level inputs are an important feature. This saves you from having to buy a line output converter to convert your head unit's speaker-level signals to preamp level. Preamp-level inputs are provided as standard equipment on almost every crossover. If you plan to use them, buy a model with ground loop isolation. This will go a long way toward preventing system noise problems.

Output level controls are usually provided for each set of outputs, but they are of limited value. They are usually redundant with the input level controls found on your amp.

There are many names for automatic amp wake-up output, a useful feature that eliminates having to run a remote turn-on wire from your head unit to your sub amp. It works by monitoring the DC voltage on a speaker wire from the head unit, and provides a signal to activate the subwoofer amp when the head unit is on. This can be a big time-saver when you're adding a subwoofer system in the trunk by tapping into the rear speaker wires and you have no other reason to pull the head unit out.

One final point. The important thing is to obtain the performance you need in your combined system. Many of the features listed above may be included in your power amp. If your amp has a subsonic filter, for example, you don't need to have one in the crossover too. If neither do, you can add one externally if you want to.

Choosing a Subwoofer Amplifier

Most of what you should know about choosing amps is contained in Chap. 6. That chapter also contains information on installing amps, amplifier power and ground wiring, remote turn-on hookup, and speaker wiring. There are two additional points worth covering here: how much power you need in a subwoofer amp and the best amplifier/subwoofer configurations.

How Much Power Do You Need?

The appropriate power for a subwoofer amp mainly depends on how much power is used for the rest of the system and the subwoofer crossover frequency used.

The typical frequency distribution of power in music is shown in Table 4-4.

Notice that less than half the power in music is above 300 Hz. This means that at least half the power in music is below 300 Hz. To make sure you're covered, use the following rule of thumb:

> **Rule of Thumb**: The sub amp power should equal 1.5 times the power to the front speakers.

For example, suppose you are using a 25W × 4 amp to drive your main speakers. The rule of thumb says you should use a sub amp that can provide roughly 1.5 × (25 watts + 25 watts) = 75 watts to your subs. This could mean a 35W × 2 amp driving a dual voice coil sub or an amp that produces 75 watts when bridged to a single sub.

		Maximum Power Above
TABLE 4-4	**Frequency**	**That Frequency**
Frequency Distribution of Power in Music	300 Hz	50%
	600 Hz	25%
	1200 Hz	10%
	2400 Hz	5%

This rule of thumb is quite generous with power to the sub for typical situations, but not all situations are typical. Consider using more power for your subwoofer amp if:

- Your subwoofer crossover frequency is higher than 150 Hz.
- Your subwoofer sensitivity is exceptionally low.
- Your subwoofer is installed in a trunk that is sonically isolated from the passenger compartment.
- You like music with lots of deep bass.

The Best Amplifier/Subwoofer Configurations

You should choose the right amplifier configuration to provide efficient power transfer to your speakers based on the number of subwoofers you plan to use and their impedances. Only by doing this are you getting the amplifier power you paid for.

Subwoofers are commonly available in 4 and 8 ohms, with single or dual voice coils. Amplifiers may or may not support bridging and may or may not be 2-ohm stable. Because of the many possible combinations, the best configurations are shown for convenience in Table 4-5.

Table 4-5 shows maximum power and rated power configurations for both one- and two-subwoofer systems. Rated power configurations provide acceptably efficient power transfer to your speakers with little risk of overheated amplifiers. Rated power configurations are so named because they give you the rated power nominally specified using a 4-ohm load. Maximum power configurations provide the maximum power transfer to your speakers that is still safe for your amp. Maximum power configurations typically provide 50 percent more power transfer than rated power configurations for a given amp.

Tip: An amp used in a maximum power configuration runs hotter, so be sure to mount it where it will have good air circulation.

TABLE 4-5

Amplifier/
Subwoofer
Configurations

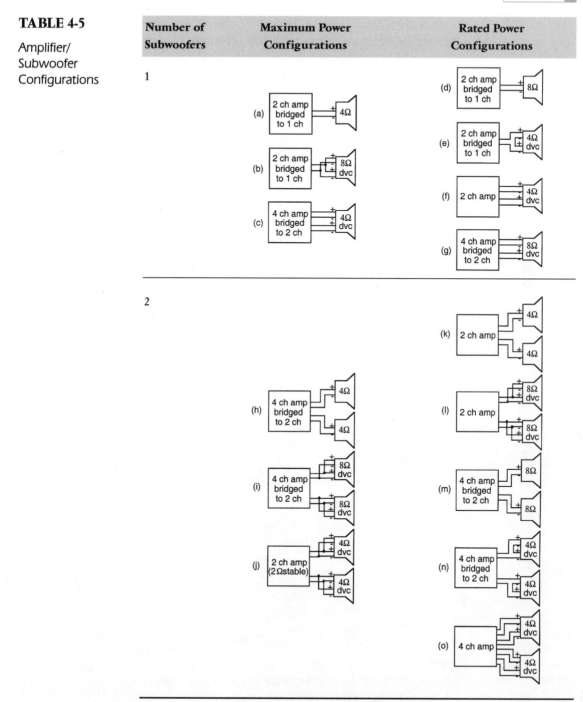

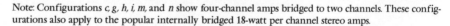

Note: Configurations *c, g, h, i, m,* and *n* show four-channel amps bridged to two channels. These configurations also apply to the popular internally bridged 18-watt per channel stereo amps.

Configurations other than those shown may not provide satisfactory power to the speakers or may risk blowing the amp. For example, using rated power configurations with higher-impedance speakers than those shown is throwing away amplifier horsepower. Using maximum power configurations with lower-impedance speakers than those shown will most likely overheat your amp.

Bridging

Bridging converts a stereo amp into a mono amp, which gives you more power than the sum of the two stereo channels. Of course, you must use an amp that is bridgeable to do this. As an example, a 50-watt per channel stereo amp might be bridgeable to a 150-watt mono configuration. Similarly, a 50-watt per channel four-channel amp might be bridgeable to a 150-watt per channel stereo configuration.

You must use a 4-ohm or higher speaker load when bridging. This means you can't run two 4-ohm woofers wired in parallel, for example. This is because a bridged amplifier "sees" a speaker load of half the actual value, so a 4-ohm speaker looks like a 2-ohm equivalent load. Very few amps can handle using anything less than 4 ohms in the bridged mode.

Be sure to follow the instructions included with the amp for the correct bridging procedure and connections—not all amps are alike.

Series and Parallel Wiring

If you want to try your hand at creating your own amplifier/speaker configurations, you will need to calculate the combined impedance of series- and parallel-wired speakers.

Series wiring refers to connecting multiple speakers (or the two sets of terminals on dual voice coil woofers) as shown in Fig. 4-1.

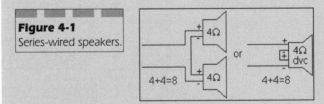

Figure 4-1
Series-wired speakers.

Notice that the plus terminal of one speaker (or set of terminals) is always connected to the minus terminal of the other. This is so both speakers (or voice coils) are driven in phase with each other. Calculating the combined impedance of series-wired speakers is easy—just add up the individual impedances. This is true for any number of speakers.

Parallel wiring refers to connecting multiple speakers (or the two sets of terminals on dual voice coil woofers) as shown in Fig. 4-2.

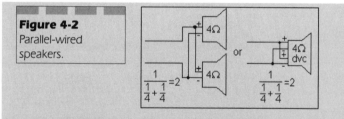

Figure 4-2
Parallel-wired speakers.

Notice that the plus terminal of one speaker (or set of terminals) is always connected to the plus terminal of the other. This is so both speakers (or voice coils) are driven in phase with each other. Calculating the combined impedance of parallel-wired speakers is more complicated. You need to add up 1 over each of the individual impedances, then take 1 over the total. This formula works for any number of speakers.

Once you understand the two basic calculations, you can calculate the impedance of combined series and parallel configurations. All you need to do is break the problem down by calculating simple series or parallel pieces of the configuration first. By substituting in the calculated values for these pieces, you'll be able to calculate the next level up until you're done.

The key to maximizing power in amplifier/subwoofer configurations is to present each amplifier with the lowest impedance load it can safely handle. This is usually 2 ohms for nonbridged amps and 4 ohms for bridged amps, but you should check the specs of any amp you plan to use.

You should avoid series wiring of speakers, with two exceptions—series wiring of the two voice coils of a dual voice coil subwoofer and series wiring of two opposing drivers mounted in an isobarik system. Wiring drivers in series otherwise can cause the back EMF of one driver to modulate the other driver, increasing distortion.

Subwoofer Enclosure Options

When it comes to choosing a subwoofer enclosure, there are two factors involved—the amount of assembly you prefer and the type of enclosure that is best for your installation. These two factors are interrelated, since the type of enclosure you choose might depend on the amount of assembly you prefer, and the amount of assembly you prefer might depend on the type of enclosure.

Subwoofer Amount of Assembly Options

If you plan to use one of the powered subwoofer configurations, then your subwoofer and box are already designed and assembled for you. Otherwise you have the flexibility to choose the amount of assembly you prefer. The three possibilities are shown in Table 4-6.

TABLE 4-6

Subwoofer Amount
of Assembly
Options

Amount of Assembly	Comments
Ready-made subwoofer	The easiest approach, and sure to look professional. No guarantee of good sound quality.
Component woofers in a prefab box	Gives you the flexibility to choose both enclosure and drivers without having to build the box yourself. Can save money compared to ready-made and provide better performance.
Build your own box	If you have basic woodworking skills, can save a lot of money. Lets you create any size and shape enclosure you wish, making best use of odd space.

A ready-made subwoofer is by far the easiest approach, and is sure to look professional. For people who have no interest in delving into box science or construction, this is the way to go. Unfortunately it's no guarantee of good sound quality. If you choose this route, either buy a product with a solid reputation or audition before you buy. Don't trust specs to tell the whole story—a lot of manufacturers bend the truth in the way they specify performance.

If you're willing to spend the extra money, some manufacturers such as JL Audio and Bostwick offer ready-made subwoofers custom designed for popular makes of trucks and sport utility vehicles (Fig. 4-3). The fiberglass enclosures are contoured to precisely fit behind a seat and are carpeted to match the vehicle's interior. The sound is tailored to match the vehicle's acoustics as well.

Figure 4-3
JL Audio Stealthbox.
(*Courtesy of JL Audio.*)

Installing component drivers in a prefabricated box gives you the flexibility to choose both enclosure and drivers without having to build the box yourself. If you choose your speakers wisely, you can save money compared to ready-made and get better performance. (Advice on choosing a speaker and box is given later in this chapter.) Prefabricated boxes are available in a variety of sizes, shapes, and types (sealed, ported, bandpass) to accommodate almost any size driver or drivers. They generally come carpeted and have speaker terminals installed.

Some prefabricated boxes are custom made for specific applications. An example is the Q-Customs Camaro/Firebird/Trans Am enclosure (Fig. 4-4). It's designed to fit precisely in the rear storage well and accommodates a pair of 12-inch woofers.

If you have basic woodworking skills, building your own box can save a lot of money. It also lets you create any size and shape enclosure you wish, making best use of whatever odd space is available. Box design and construction are covered later in this chapter.

Types of Enclosures

There are many types of speaker enclosures you can choose from, including sealed, ported, bandpass, and transmission line. Whether you plan to buy a ready-made subwoofer, use a prefab box, or build your own, it's important that you understand the fundamental differences.

If you've ever seen a pair of 6 × 9 speakers wedged between the rear window and deck of a car, with no enclosure, you may have wondered: Why use a box at all? The basic answer is that the back of a speaker makes as much sound as the front, but of the opposite polarity. If you happen to be sitting where you get an equal amount of sound from front and

Figure 4-4
Q-Customs
Camaro/Firebird
enclosure—(a) before;
(b) after. (*Courtesy of Ai
Research.*)

(a)   (b)

back, the two will totally cancel each other out and you'll hear nothing! In practice, speakers without any enclosure will provide some high frequencies and mid-range, but no deep bass. This is because the sound from the back of the speaker gets delayed due to a longer reflected path to your ear. This causes sound cancellation at some frequencies (especially low ones) but reinforcement at others.

Using an enclosure is the practical solution to this problem. The most popular enclosure types for subwoofers are shown in Table 4-7.

Free-Air. Free-air (or enclosureless) subwoofers are designed to be mounted on the rear deck or behind the back seat, using the trunk as a large speaker enclosure. The idea here is for the trunk to isolate all of the sound from the back of the speaker, thus avoiding the problem of sound cancellation explained previously.

Assuming you have a trunk and a suitable mounting location, free-air subwoofers eliminate the need to build or buy a box. The problem is that most trunks do not provide good sonic isolation. This means that sound from the back of the speaker can be heard in the passenger compartment, canceling out deep bass. To check this for yourself, have a friend with a deep voice talk into your open trunk from the outside. You should be listening inside the passenger compartment with the windows rolled up. If you can barely hear your friend, then a free-air subwoofer will work in your car. Otherwise, choose one of the other options.

Sealed. Using a sealed box effectively prevents sound from the back of the speaker from being heard, but also affects the sound coming from the front of the speaker. The air inside a sealed box acts like a spring, effectively changing the behavior of the speaker. How this affects the sound depends on the size of the box and the particular driver.

Sealed boxes offer a number of advantages over the other enclosure types. They are the easiest box type to design and build. Sealed systems are the least sensitive to using a box that is a little too big or small, or to drivers that are different than their published specs. This is a big benefit if you are planning to use component woofers in a prefab box. The spring effect of air in a sealed box provides woofer protection against turn-on thumps, subsonic bass energy, and going over bumps.

Sealed boxes provide the deepest bass if you must use a small enclosure. The big drawback of sealed boxes is that they cannot provide bass as deep as the other enclosure types for larger enclosures. However, the low-frequency rolloff characteristic of sealed systems is relatively shallow compared with that for ported and bandpass systems, resulting in useful deep bass below the "official" −3-dB cutoff frequency. Exactly what defines a

TABLE 4-7

The Most Popular
Enclosure Types for
Subwoofers

Enclosure Type	Diagram	Comments
Free-air (trunk is enclosure)	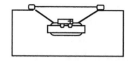	Requires a trunk as the enclosure. Eliminates the need to build or buy a box. Poor performance in most cases due to poor sonic isolation of trunk causing deep bass cancellation.
Sealed	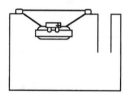	Easiest box type to design and build. Woofer is protected against turn-on thumps and subsonic bass. Provides the deepest bass if you must use a small enclosure. Cannot provide bass as deep as the other enclosure types for large enclosures.
Ported	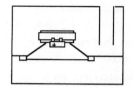	More difficult to design and build than sealed systems. Provides no woofer protection against turn-on thumps and subsonic bass. Can provide bass at least an octave deeper than sealed boxes for a given driver. Requires a much bigger box than sealed systems.
Bandpass	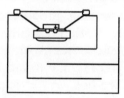	More difficult to design and build than either sealed or ported systems. Produces sound over a narrow frequency range, reducing the requirement for a steep low-pass slope in a subwoofer crossover. Runs the risk of sounding like a "one-note" system. Requires a much bigger box than sealed systems.
Transmission line		Elaborate and expensive construction required to create the internal labyrinth structure. Possible to achieve deeper bass than with a sealed box. Not recommended.

small enclosure and how deep a sealed box can go is covered in the "Box Design" section later in this chapter.

Ported.   Ported enclosures make constructive use of the sound from the back of the speaker. Because of this, it becomes possible to achieve deeper bass than with a sealed box. In a ported system, the sound from the back of the speaker produces a resonant output at the port, much like producing a tone by blowing air across the mouth of a soda bottle. The port is tuned to provide constructive reinforcement of bass at the proper frequency.

Ported systems are more difficult to design and build than sealed systems. They are more sensitive than sealed systems to using a box that is a little too big or small and to drivers that are different than their published specs. Ported boxes provide no woofer protection against turn-on thumps, subsonic bass energy, and going over bumps.

The big plus of ported boxes is that they can provide bass at least an octave deeper than sealed boxes. That's the difference between 80-Hz bass and 40-Hz bass. However, a much bigger box—typically five times bigger—is required to get that octave improvement.

A primary factor in whether to use a ported box or a sealed box is how big a box you can tolerate. Small boxes should always be sealed, and large boxes should generally be ported. What is considered a big or small box is explained in the "Box Design" section later in this chapter.

Bandpass.   Bandpass enclosures are based on ported enclosures, except that the direct sound from the speaker is blocked off using a secondary sealed chamber. This means all you hear is the tuned output from the port. Unlike the other enclosure types, bandpass boxes only pass sound over a narrow frequency band. This reduces the requirement for a steep low-pass slope in a subwoofer crossover. Like the tone from a soda bottle, however, bandpass designs run the risk of sounding like "one-note" systems.

Bandpass systems are more difficult to design and build than either sealed or ported systems. They are very sensitive to speaker parameters, box volumes of the sealed and ported sub-enclosures, and port tuning. Because of this, using component woofers in a "one size fits all" prefab bandpass box is insanity.

Bandpass systems produce deep bass that is comparable to that of equally sized ported systems. Like ported systems, bandpass systems require big enclosures. A unique property of bandpass systems is that they allow you to trade efficiency for bandwidth by changing the sealed fraction of the box. Very efficient bandpass designs have inherently narrow bandwidths, resulting in a "one-note thump monster." The efficiency/ bandwidth tradeoff is shown in detail in the "Box Design" section later in this chapter.

Transmission Lines.   Even though the way they work is different than ported enclosures, transmission line enclosures also make constructive use of the sound from the back of the speaker. Doing so makes it possible to achieve deeper bass than with a sealed box.

Transmission line enclosures delay the sound from the back of the speaker through a labyrinth structure. The delay is chosen to provide constructive reinforcement of bass at the proper frequency.

Transmission line enclosures utilize elaborate and expensive construction to create the internal labyrinth structure. They do not offer any advantages over properly designed ported enclosures and often have coloration problems in the mid-bass region. Their design typically involves more trial and error than science to produce a satisfactory result. For these reasons, transmission line enclosures are not recommended.

Choosing a Driver

Whether you buy a pre-made box or build your own, you will need to choose your own subwoofer driver.

Choosing a driver and buying or building a box are interrelated because your choice of driver affects the size and type of box required. Most people choose a driver, then buy or build a suitable box. The problem is that you can end up needing a box that is bigger than you have space for or not the type you want. This section will help you avoid that problem by showing you how you can quickly and easily estimate what box you would need for a specific driver. It will also cover power handling, sensitivity, and other factors influencing your choice of driver.

Power Handling

How much power a subwoofer driver can handle is limited by two things— the diameter and maximum excursion of the cone and the power rating of the voice coil. Either factor can be the weak link in power handling.

At low frequencies, a subwoofer has to move lots of air to create loud bass. How much air it can move is proportional to the area of the cone multiplied by the maximum excursion of the cone. Since area is proportional to cone diameter squared, large drivers have the potential to move a lot more air than small ones. Cone excursion means how far out the cone can realistically move before distortion or damage to the speaker occurs. This is shown as Xmax on subwoofer spec sheets.

Voice coil power handling depends on the diameter of the voice coil, the gauge of wire used, and other factors affecting how much heat the voice coil assembly can handle and effectively dissipate.

Tip: Be careful when reading manufacturers' literature—some list Xmax as total excursion, others as one-way excursion, still others as one-way + 15%!

When choosing a driver, make sure its power rating exceeds that of the amplifier you'll be driving it with. Unfortunately, many manufacturers make exorbitant claims about power handling for their subwoofers. If it seems too good to be true, compare the Xmax and voice coil diameter specs against a reputable driver with the same size cone. If the driver in question claims higher power handling with similar specs or similar power handling with a smaller Xmax or voice coil diameter, watch out.

Sensitivity

Sensitivity tells you how loud a speaker will play at a specified power level. This number is given in decibels and is normally measured at 1 meter using 1 watt. A typical value is 90 dB, and higher numbers are better.

For example, a subwoofer with 90-dB sensitivity is 3 dB better than one with 87 dB. This means that it is 3 dB louder—equivalently, it would require 3 dB less amplifier power to play at the same volume level. Table 4-8 illustrates how sensitivity affects how much power you would need to produce the same volume level.

Table 4-8 shows that the 87-dB driver would require 200 watts of power to play as loud as the 90-dB driver using 100 watts. This seems dramatic, but you may not achieve the big benefit you expect, since high-sensitivity drivers often produce higher cutoff frequencies. This means that the gains you pick up in sensitivity may be lost in having to boost the deep bass frequencies.

The bottom line is not to overemphasize the importance of sensitivity. Consider it along with bass response to paint a more complete picture.

Bass Response

The bass response of a subwoofer system depends on two things—the driver itself and its enclosure. In general, the larger the enclosure you are willing to use, the lower the bass response will be for a properly designed

TABLE 4-8

How Sensitivity
Affects Required
Power

Sensitivity	Required Amp Power
87 dB	200 W
88 dB	158 W
89 dB	126 W
90 dB	100 W
91 dB	79 W
92 dB	63 W
93 dB	50 W

sealed, ported, or bandpass system. For free-air applications, bass response depends only on the driver.

The most accurate way to compare the bass response potential of various drivers is to simulate the frequency response of each in your intended application. This would entail a large amount of work, not to mention the fact that you may not have even decided what size your intended enclosure should be.

However, you don't need to go through all that work. You can calculate a single number for each of the drivers you're considering that lets you compare its bass response potential to that of the others. Better yet, you can do this without knowing anything about your intended application, other than whether it's free-air or not.

Introduction to T/S Parameters. When you want to compare the bass response potential of various drivers (or, later, design a subwoofer box), there are three magic numbers that tell you everything you need to know about a driver. These are called the Thiele/Small (T/S) parameters.

T/S parameters are normally provided in mail-order catalogs or printed right on the box in stores. If you can't find them, ask the retailer—you can't do anything without them. Table 4-9 lists the T/S parameters:

The problem with T/S parameters is that they don't tell you much in their raw form. You need to use your calculator or computer to do something useful with them.

Comparing Woofers for Sealed, Ported, and Bandpass Applications. To compare woofers for sealed, ported, and bandpass applications, here's the calculation you need to perform:

$$f_{fb} = f_s \sqrt{V_{as}}$$

TABLE 4-9

Thiele/Small
Parameters

T/S Parameter	Units	Meaning
f_s	Hz	Resonant frequency of driver
Q_{ts}		Total Q at the resonant frequency
V_{as}	ft³*	Volume of air having the same compliance as driver

*Some manufacturers provide V_{as} in liters rather than ft³. Divide the number of liters by 28.3 to convert to ft³.

f_{fb} indicates how low the bass response would be for that driver in a reference-size evaluation box. (Lower numbers are better.) Suppose you are trying to decide between the two 10-inch subwoofers in Table 4-10.

Calculating f_{fb} for each driver provides the results given in Table 4-11.

This shows that the Jensen driver provides bass response about 10 percent lower than that of the Ultimate driver. (This would come as a big surprise if you used f_s to judge bass potential.) This holds true regardless of what the box volume is, so long as you are comparing both drivers using the same box volume. This simple calculation lets you compare woofers without knowing the size or type of your box.

If you know your box size and type, the actual cutoff frequencies for sealed and ported systems can be estimated using these equations:

sealed box: $\quad f_3 = 0.8 f_{fb}/\sqrt{V_b}$

ported box: $\quad f_3 = 1.0 f_{fb}/\sqrt{V_b}$

where f_3 is the cutoff frequency and V_b is the box volume in ft³.

TABLE 4-10

T/S Parameters of
Two Contending
Subwoofers

Model	f_s	Q_{ts}	V_{as}
Ultimate AU1050	29 Hz	0.43	3.50 ft³
Jensen JSW104	31.3 Hz	0.40	2.40 ft³

TABLE 4-11

Relative Bass
Response of Two
Contending Sub-
woofers

Model	f_{fb}
Ultimate AU1050	54.3
Jensen JSW104	48.5

Tip: Avoid drivers with
unusually high Q_{ts} values—anything above
0.8 is too high. High
values are caused by
undersized magnet
structures and usually
result in frequency
responses with a large
resonant peak.

These equations show that a larger box provides a lower cutoff frequency. Furthermore, it takes a box that is 4 times larger to lower the cutoff frequency by a factor of 2. Suppose you have space for a 0.75-ft³ box behind the seat of your truck and you have chosen to use a sealed box. The approximate cutoff frequencies for the above subwoofers would be as shown in Table 4-12.

Comparing Woofers for Free-Air Applications. To compare woofers for free-air applications, the calculation is different:

$$f_{ob} = f_s / Q_{ts}$$

f_{ob} indicates how low the relative bass response would be in a gigantic box. (Lower numbers are better.) Calculating f_{ob} for the same two 10-inch subwoofers used previously provides the results shown in Table 4-13.

In this case, the Ultimate driver is better—it provides a relative cutoff frequency about 15 percent lower than that of the Jensen driver. The actual cutoff frequencies can be estimated using:

$$f_3 = 0.9 \ f_{ob}$$

Calculating for the previously mentioned subwoofers gives us the values in Table 4-14.

Notice that these cutoff frequencies are actually worse (higher) than those obtained using the 0.75-ft³ sealed box. This demonstrates that free-air applications may sacrifice deep bass compared to properly designed sealed and ported systems. This is true even for the best of trunks, where sonic isolation is not an issue.

TABLE 4-12

Bass Response
Comparison in a
0.75-ft³ Sealed Box

Model	Cutoff Frequency in 0.75-ft³ Sealed Box
Ultimate AU1050	50.1 Hz
Jensen JSW104	44.8 Hz

TABLE 4-13

Relative Free-Air
Response
Comparison

Model	f_{ob}
Ultimate AU1050	67.4
Jensen JSW104	78.3

TABLE 4-14

Actual Free-Air
Response
Comparison

Model	Free-Air Cutoff Frequency
Ultimate AU1050	60.7 Hz
Jensen JSW104	70.5 Hz

Cone and Surround Material

Cone material (Table 4-15) and surround material (Table 4-16) are important in the harsh automotive environment, but it's hard to wade through the hype. Polypropylene cones are extremely resistant to environmental deterioration, but are a temperature-sensitive plastic and become soft in the heat, hard when cold. Paper cones deteriorate with sunlight and moisture, so you must keep them out of the sun. Coated paper is a good compromise. Other materials such as graphite, Tri-Laminate, resin laminate, carbon-blended poly, kapok, poly-graphite, graphite-quartz, Foam-Infused IMPP, titanium composite, fiberglass, and Kevlar offer the promise of superior performance. Don't expect a big improvement over coated paper.

The surround is the soft ring around the outside of the cone with the bulge in it. Foam deteriorates with sunlight—choose rubber if your subwoofer will be exposed to the sun.

Dual Voice Coil Subwoofers

A dual voice coil speaker (Fig. 4-5) is simply one in which two separate voice coil windings and sets of terminals are provided. It's more expensive to wind and terminate dual voice coils, but you typically pay only a small premium compared to a similar single voice coil speaker.

TABLE 4-15

Common
Subwoofer Cone
Materials

Cone Material	Comments
Paper	Deteriorates with sunlight and moisture
Coated paper	Good choice
Polypropylene	Temperature-sensitive performance

TABLE 4-16

Common Sub-
woofer Surround
Materials

Surround Material	Comments
Foam	Deteriorates with sunlight
Butyl rubber	Good choice

The main advantage of dual voice coil speakers is wiring flexibility. A single dual voice coil driver offers the user three hookup choices: parallel, series, and independent. In a parallel hookup the drivers' impedance will be half that of each individual coil (a dual 4-ohm speaker would be a 2-ohm speaker in parallel). A series hookup results in twice the impedance of each single coil (a dual 4-ohm speaker results in 8 ohms if its coils are wired in series). Finally, you can wire each voice coil to a separate channel of your amplifier, which can be useful if your amplifier is not mono-bridgeable or if you are bridging a four-channel amplifier down to two channels to run your sub.

If a dual voice coil subwoofer is wired to two independent channels, the inputs to both channels should ideally be the same (summed mono). If there is any difference between the signals driving each coil, the voice coils effectively fight each other. This results in wasted amplifier power, reduced headroom, and increased distortion.

A common misconception with regard to dual voice coil speakers is the assumption that nothing changes if you use only one of the voice coils. With only one coil hooked up, a dual voice coil speaker will suffer a loss in reference efficiency of about 3 dB as well as a significant shift in its Thiele/Small parameters (Q_{ts} will go way up). This renders any enclosure calculations inaccurate unless you measure the speaker parameters with only one coil hooked up. Failure to account for the different parameters

Figure 4-5
Madisound
12204DVC sub-
woofer. (*Courtesy of
Madisound.*)

Tip: For independently wired DVC subs, adding a choke coil in series with each winding of the subwoofer (Fig. 4-6) will prevent power amp oscillation problems. The interwinding capacitance or mutual inductance of DVC subwoofers can cause some power amplifiers to oscillate. The oscillation can be at almost any frequency, from very low ("motorboating") to very high—even ultrasonic. Ultrasonic oscillation is not directly audible, but results in reduced headroom and increased distortion. In some cases, oscillation can result in amplifier overheating or destruction.

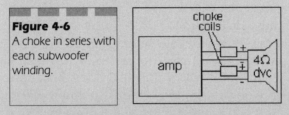

Figure 4-6
A choke in series with each subwoofer winding.

You can purchase choke coils inexpensively or wind your own. Twenty turns of #22 wire around a ¼-inch bolt (keep the bolt in) will provide about 20 μH of inductance—a good value. Make sure the DC resistance is less than 0.1 ohms to prevent signal loss to the subwoofer and ensure adequate power handling of the choke.

of a dual voice coil speaker with only one coil powered can result in very poor performance.

Box Design

Whether you buy a pre-made box or build your own, this section will show you how to design or choose the right box for your driver.

Computer Programs for Box Design

There are many excellent computer programs for helping you design a speaker box. Many of them are available at little or no cost. *The Loudspeaker Design Cookbook* by Vance Dickason lists and describes the most well known of these programs.

Even if you plan to use a computer program, this section will give you an intuitive understanding of how box size and enclosure type affect the bass response. If you don't have a computer, this section has everything you need to design your enclosure.

Subwoofer Design Toolbox

The Subwoofer Design Toolbox by MFR Engineering (Fig. 4-7) is an easy-to-use but powerful program for designing subwoofers in Microsoft Windows (3.1 or higher). The program's convenient tab interface lets you choose from box design, port design, or woofer selection tools.

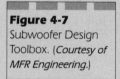

Figure 4-7
Subwoofer Design Toolbox. (*Courtesy of MFR Engineering.*)

Box designs can be sealed, ported, or bandpass. Context-sensitive design hints and recommended values assist beginners and pros alike. The woofer selection tool lets you quickly compare woofers without having to design a box for each of them.

With the Subwoofer Design Toolbox, it's easy to compare frequency responses of different box designs on a single graph. Frequency response printouts can be in color, or with symbols enabled to improve readability with black-and-white printers.

Features of particular interest to automotive applications include free-air modeling, multiple port design, and isobarik loading.

Step 1: Calculate f_{ob} and V_{of} from the T/S Parameters

As explained in the "Choosing a Driver" section earlier in this chapter, there are three magic numbers that tell you everything you need to know about a driver for the purposes of box design. These are called the Thiele/Small (T/S) parameters. As mentioned earlier, T/S parameters are normally provided in mail-order catalogs or printed right on the box in stores. They are listed again in Table 4-17 for convenience.

The first step in designing your box is calculating the following two parameters, plugging in the T/S parameters of your driver:

TABLE 4-17

Thiele/Small
Parameters

T/S Parameter	Units	Meaning
f_s	Hz	Resonant frequency of driver
Q_{ts}		Total Q at the resonant frequency
V_{as}	ft³*	Volume of air having the same compliance as driver

*Some manufacturers provide V_{as} in liters rather than ft³. Divide the number of liters by 28.3 to convert to ft³.

$$f_{ob} = f_s / Q_{ts}$$

$$V_{of} = V_{as} \cdot Q_{ts}^2$$

These two parameters will be used many times throughout the design process.

Step 2: Label the Frequency Response Diagram for Your Driver

The next step is to refer to the appropriate frequency response diagram. Choose the diagram with the nearest Q_{ts} value to your driver, as well as the desired enclosure type (Figs. 4-8 and 4-9). For each value of Q_{ts} from 0.2 to 0.8, there is a combined sealed and ported diagram and a separate band-pass diagram.

Once you've identified the proper diagram, it's time to label the frequency axis of the diagram using the value of f_{ob} calculated in Step 1. Relabel 0.1 f_{ob}, 1 f_{ob}, and 10 f_{ob} frequency points with their calculated values. Label intermediate points of 0.2 f_{ob} through 0.9 f_{ob} and 2 f_{ob} through 9 f_{ob} for more detail. You might want to make a photocopy of the diagram, or you can write directly in the book with pencil.

Similarly, you'll need to label the enclosure volume for each curve, using the value of V_{of} calculated in Step 1.

Step 3: Choose Your Box Size and Type

Once you've labeled the curves, you'll have a graphic representation of what is possible for your driver. You can now choose the best trade-off between box size and bass response for your needs.

Figure 4-8

Figure 4-8
Frequency response
diagrams for sealed
and ported drivers.
(a) $Q_{ts} = 0.2$;
(b) $Q_{ts} = 0.3$.

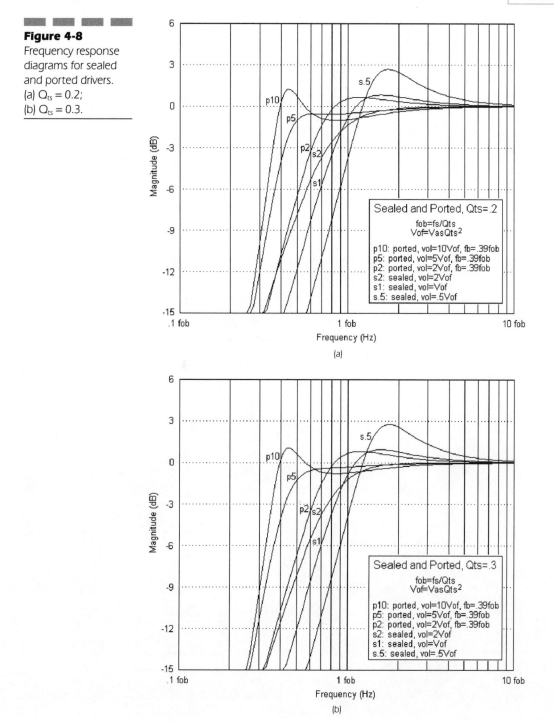

Sealed and Ported, Qts=.2

fob=fs/Qts
Vof=VasQts²

p10: ported, vol=10Vof, fb=.39fob
p5: ported, vol=5Vof, fb=.39fob
p2: ported, vol=2Vof, fb=.39fob
s2: sealed, vol=2Vof
s1: sealed, vol=Vof
s.5: sealed, vol=.5Vof

(a)

Sealed and Ported, Qts=.3

fob=fs/Qts
Vof=VasQts²

p10: ported, vol=10Vof, fb=.39fob
p5: ported, vol=5Vof, fb=.39fob
p2: ported, vol=2Vof, fb=.39fob
s2: sealed, vol=2Vof
s1: sealed, vol=Vof
s.5: sealed, vol=.5Vof

(b)

■■■ ■■■ ■■■ ■■■

**Figure 4-8
(continued)**
Frequency response
diagrams for sealed
and ported drivers.
(c) $Q_{ts} = 0.4$;
(d) $Q_{ts} = 0.5$.

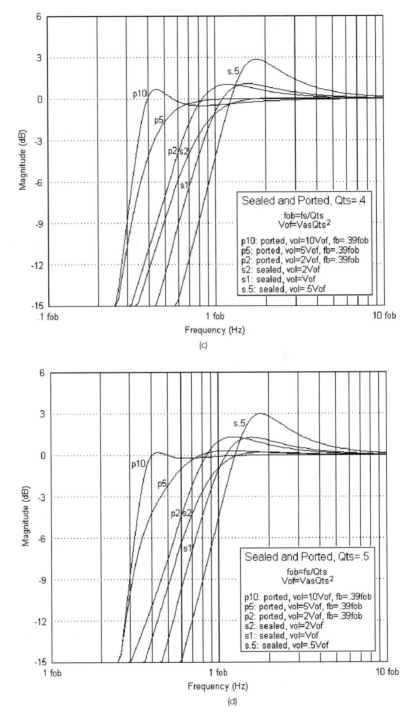

Sealed and Ported, Qts=.4

fob=fs/Qts
Vof=VasQts²

p10: ported, vol=10Vof, fb=.39fob
p5: ported, vol=5Vof, fb=.39fob
p2: ported, vol=2Vof, fb=.39fob
s2: sealed, vol=2Vof
s1: sealed, vol=Vof
s.5: sealed, vol=.5Vof

(c)

Sealed and Ported, Qts=.5

fob=fs/Qts
Vof=VasQts²

p10: ported, vol=10Vof, fb=.39fob
p5: ported, vol=5Vof, fb=.39fob
p2: ported, vol=2Vof, fb=.39fob
s2: sealed, vol=2Vof
s1: sealed, vol=Vof
s.5: sealed, vol=.5Vof

(d)

**Figure 4-8
(continued)**
Frequency response
diagrams for sealed
and ported drivers.
(e) $Q_{ts} = 0.6$;
(f) $Q_{ts} = 0.7$.

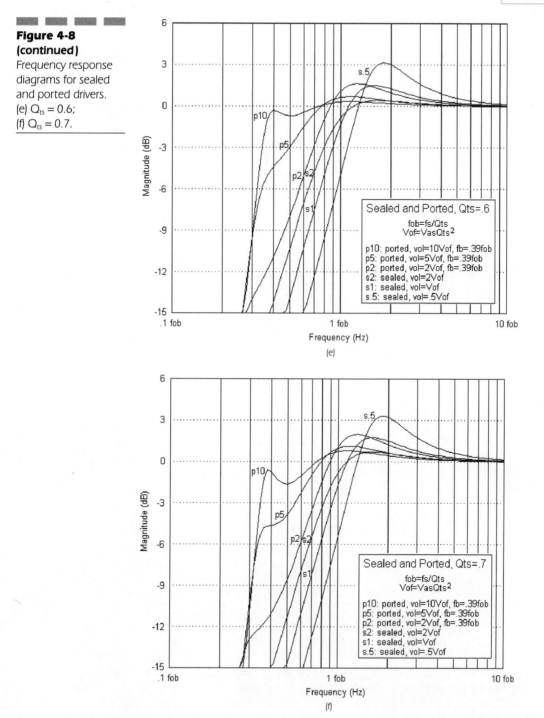

(e)

(f)

**Figure 4-8
(continued)**

Frequency response
diagrams for sealed
and ported drivers.
(g) $Q_{ts} = 0.8$.

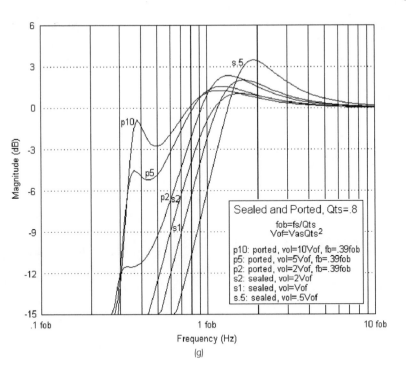

Sealed and Ported, Qts=.8

fob=fs/Qts
Vof=VasQts²

p10: ported, vol=10Vof, fb=.39fob
p5: ported, vol=5Vof, fb=.39fob
p2: ported, vol=2Vof, fb=.39fob
s2: sealed, vol=2Vof
s1: sealed, vol=Vof
s.5: sealed, vol=.5Vof

(g)

Figure 4-9

Frequency response
diagrams for band-
pass drivers.
(a) $Q_{ts} = 0.2$

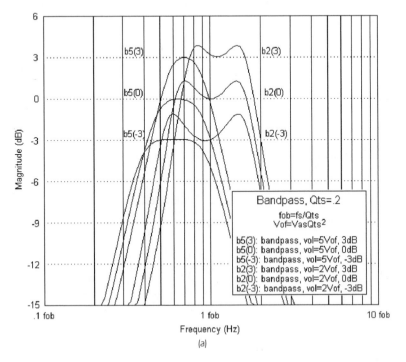

Bandpass, Qts=.2

fob=fs/Qts
Vof=VasQts²

b5(3): bandpass, vol=5Vof, 3dB
b5(0): bandpass, vol=5Vof, 0dB
b5(-3): bandpass, vol=5Vof, -3dB
b2(3): bandpass, vol=2Vof, 3dB
b2(0): bandpass, vol=2Vof, 0dB
b2(-3): bandpass, vol=2Vof, -3dB

(a)

**Figure 4-9
(continued)**
Frequency response
diagrams for band-
pass drivers.
(b) $Q_{ts} = 0.3$;
(c) $Q_{ts} = 0.4$.

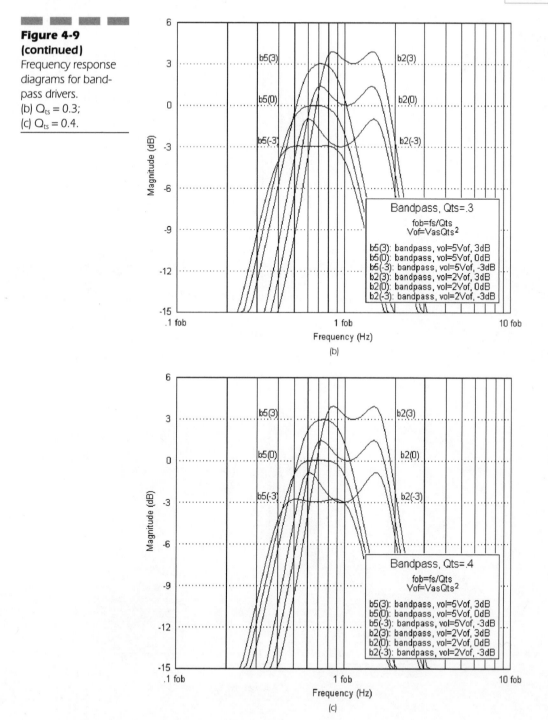

**Figure 4-9
(continued)**
Frequency response
diagrams for band-
pass drivers.
(b) $Q_{ts} = 0.3$;
(c) $Q_{ts} = 0.4$.

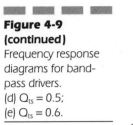

**Figure 4-9
(continued)**

Frequency response
diagrams for band-
pass drivers.
(d) $Q_{ts} = 0.5$;
(e) $Q_{ts} = 0.6$.

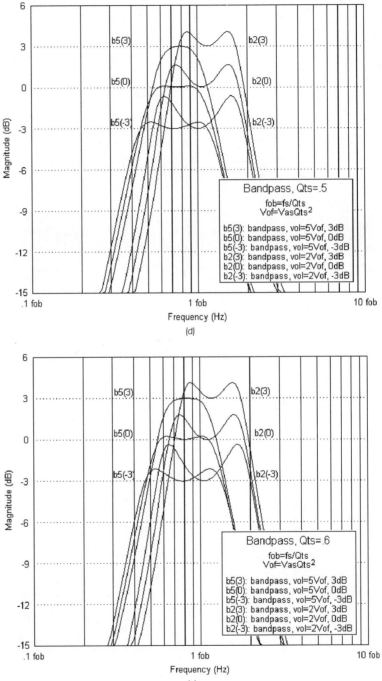

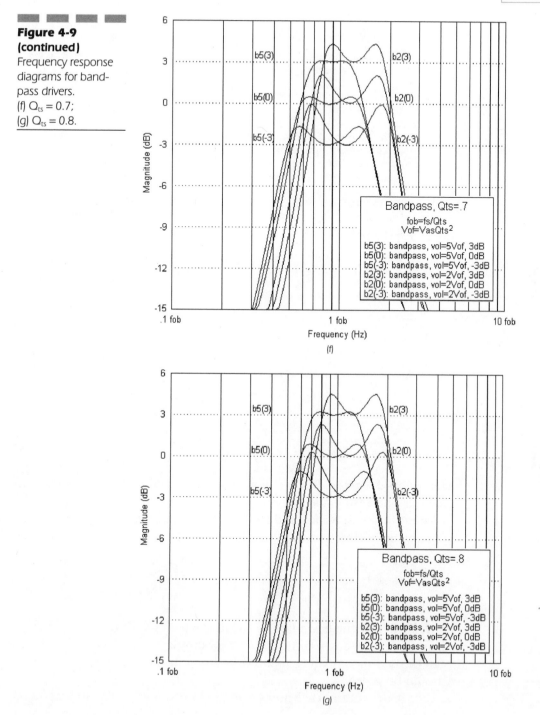

Figure 4-9
(continued)
Frequency response
diagrams for band-
pass drivers.
(f) $Q_{ts} = 0.7$;
(g) $Q_{ts} = 0.8$.

How Low Do You Need to Go?

Deciding how low in frequency you want your subwoofer to go is usually a compromise between performance and practical box size. It's easy to weigh the acceptability of a particular box size, but how do you weigh the importance of bass response?

Today's car cassette players have bass responses anywhere from 50 Hz down to about 30 Hz. Similarly, FM radio broadcasts are limited to between 30 and 50 Hz. If you have a CD player, you can expect response down to 20 Hz or even lower. This doesn't mean your music goes that low, it just means that your CD player can pass it if it does.

In spite of the fact that humans can hear (or feel) bass below 20 Hz, 30 Hz is the realistic lower limit of recorded music. The truth is, very little recorded music (including CDs) contains much below about 40 Hz. Between 40 and 60 Hz, kick drums, tympani, and the low incisive "chunk" of bass guitars kick in. Between 60 and 100 Hz is where you find the power of drums and the deep, tight, strong bass that makes rock solid.

Table 4-18 gives you some idea of what to shoot for.

TABLE 4-18

Subwoofer Response Goal

Subwoofer Response	Comment
30 Hz	Overkill
40 Hz	Realistic goal for auto sound
50 Hz	Respectable performance
60 Hz	Acceptable performance

Sealed and Ported Boxes. For both sealed and ported boxes, you can see that increasing the box size results in deeper bass. Notice that the smallest boxes are sealed and the largest ones are ported. This is because porting helps large boxes, but hurts small ones.

Rule: Use a port when the box volume is greater than 2 V_{of}.

2 V_{of} defines the boundary between small and large boxes for a particular driver. For box volumes near this boundary, sealed or ported boxes can be used successfully. Larger boxes should be ported and smaller boxes should be sealed.

For sealed and ported boxes, you are not limited to the six curves on each diagram. You can easily use a box size that is between those shown.

In that case, your results will be between the nearest two curves. You can also use a box that is slightly smaller or larger than any of those shown, with predictable results.

Bandpass Boxes.   As with sealed and ported boxes, increasing the size of bandpass boxes results in deeper bass. With bandpass boxes, however, you have an additional degree of flexibility—you can trade sensitivity for bandwidth by changing how much of the total box volume is used for the ported section.

There are six curves shown for each of the bandpass diagrams. The three to the left use a box volume of 5 V_{of}, and the three to the right use a box volume of 2 V_{of}. For each of the two box sizes there are three curves: 3 dB, 0 dB, and −3 dB. (0 dB on these diagrams is identical to 0 dB on the sealed and ported diagrams, so you can directly compare them.)

The 3-dB bandpass designs play 3 dB louder than the 0-dB designs (or require half the amplifier power to play at the same level). The price you pay for using them is their narrow bandwidth. Narrow bandwidths can leave a hole in the system frequency response if the subwoofer's high-end response doesn't extend high enough to cover the low-end response of the main speakers. Narrow bandwidths can also cause a "one-note thump" subwoofer sound.

For bandpass boxes, you must choose one of the six curves shown on each diagram. If you want a different box volume or sensitivity, you'll need to use a computer program to design it.

Step 4: Calculate the Port Tuning Frequency

Once you've decided on a curve for your driver, you'll need to calculate the port tuning frequency. This applies only to ported or bandpass boxes—no additional calculations are needed for sealed boxes.

The port tuning frequency is normally referred to as the box frequency. Its symbol is f_b. The box frequency will be needed for port construction.

Ported Boxes.   For ported boxes, calculating the box frequency is easy. Here's the formula:

$$f_b = 0.39 \ f_{ob}$$

Bandpass Boxes.   For bandpass boxes, refer to Table 4-19 or Table 4-20 for the value of f_b/f_{ob} that applies to your design. (You will also want to make note of the ported fraction of volume entry from the table. This tells you how to partition your box internally.) To calculate the box frequency, plug the value of f_b/f_{ob} from the table and the f_{ob} of your driver into this formula:

$$f_b = \frac{f_b}{f_{ob}} f_{ob}$$

Sealed/Ported Box Design Example

For this example we'll design a box for the Madisound 12204DVC 12-inch subwoofer. The T/S parameters for this driver are listed in Table 4-21.

Calculating f_{ob} and V_{of} from the T/S parameters, we obtain:

$$f_{ob} = f_s/Q_{ts} = 69.6 \text{ Hz}$$

$$V_{of} = V_{as} \cdot Q_{ts}^2 = 0.727 \text{ ft}^3$$

Since Q_{ts} is 0.316, we will use the $Q_{ts} = 0.3$ sealed/ported diagram. The frequency axis of this diagram can now be relabeled based on $f_{ob} = 69.6$ Hz. Similarly, the enclosure volumes can now be relabeled based on $V_{of} = 0.727$ ft^3. The result is shown in Fig. 4-10.

TABLE 4-19

Bandpass
Design for Box
Volume = 2 V$_{of}$

| | +3 dB | | 0 dB | | −3 dB | |
| | Ported Fraction of | | Ported Fraction of | | Ported Fraction of | |
Q_{ts}	Volume	f_b/f_{ob}	Volume	f_b/f_{ob}	Volume	f_b/f_{ob}
0.2	0.578	1.107	0.490	1.010	0.402	0.936
0.3	0.568	1.117	0.478	1.024	0.390	0.954
0.4	0.552	1.130	0.460	1.042	0.370	0.977
0.5	0.535	1.151	0.440	1.069	0.348	1.008
0.6	0.510	1.175	0.412	1.100	0.322	1.048
0.7	0.485	1.209	0.385	1.142	0.295	1.095
0.8	0.455	1.248	0.355	1.190	0.268	1.150

TABLE 4-20

Bandpass
Design for Box
Volume = 5 V_{of}

| | +3 dB | | | 0 dB | | | −3 dB | | |
Q_{ts}	Ported Fraction of Volume	f_b/f_{ob}		Ported Fraction of Volume	f_b/f_{ob}		Ported Fraction of Volume	f_b/f_{ob}
0.2	0.565	0.707		0.475	0.649		0.388	0.606
0.3	0.540	0.724		0.445	0.671		0.355	0.632
0.4	0.502	0.749		0.405	0.704		0.315	0.672
0.5	0.458	0.787		0.358	0.749		0.270	0.724
0.6	0.405	0.834		0.308	0.806		0.228	0.787
0.7	0.355	0.894		0.262	0.872		0.192	0.859
0.8	0.305	0.963		0.222	0.947		0.162	0.937

A reasonable goal for bass response is −3 dB at or below 40 Hz. Looking at the curves, you can see that the p2 curve almost makes this goal and that the p5 curve exceeds it. Even the largest sealed box (s2 curve) doesn't look like a strong contender here. The p2 curve is −3 dB at about 44 Hz and the p5 curve is −3 dB at about 30 Hz. The box sizes required for these two designs are 1.45 and 3.64 ft³, respectively. Assume our application has room for something a little bigger than the p2, but not as big as the p5. A 2-ft³ box seems like a good compromise.

Since we decided on a ported design, we also need to calculate the box frequency:

$$f_b = 0.39 \ f_{ob} = 27.1 \text{ Hz}$$

The box frequency will be needed for port construction. The "Box Construction" section later in this chapter explains how to build a port.

TABLE 4-21

T/S Parameters
for Madisound
12204DVC

Model	f_s	Q_{ts}	V_{as}
Madisound 12204DVC	22 Hz	0.316	7.28 ft³

Figure 4-10
Relabeled
sealed/ported
diagram.

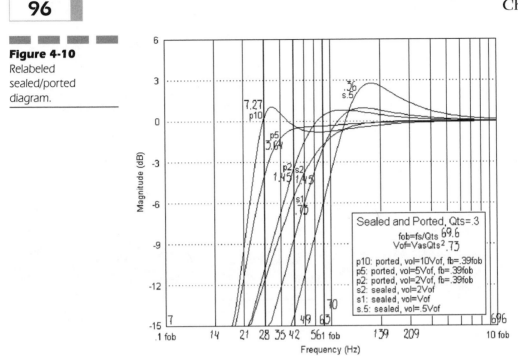

Bandpass Box Design Example

Now we'll design a bandpass box for the same driver used in the sealed/ported example. The T/S parameters for the Madisound 12204DVC 12-inch subwoofer are repeated in Table 4-22 for convenience.

Calculating f_{ob} and V_{of} from the T/S parameters, we obtain:

$$f_{ob} = f_s/Q_{ts} = 69.6 \text{ Hz}$$

$$V_{of} = V_{as} \cdot Q_{ts}^2 = 0.727 \text{ ft}^3$$

Since Q_{ts} is 0.316, we will use the $Q_{ts} = 0.3$ bandpass diagram. The frequency axis of this diagram can now be relabeled based on $f_{ob} = 69.6$ Hz. Similarly, the enclosure volumes can now be relabeled based on $V_{of} = 0.727 \text{ ft}^3$. The result is shown in Fig. 4-11.

TABLE 4-22

T/S Parameters
for Madisound
12204 DVC

Model	f_s	Q_{ts}	V_{as}
Madisound 12204DVC	22 Hz	0.316	7.28 ft^3

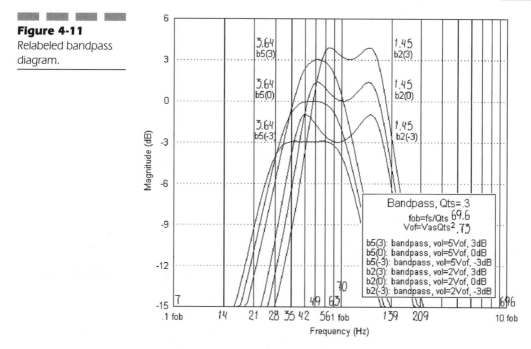

Figure 4-11
Relabeled bandpass
diagram.

We must choose from one of two box sizes here—we cannot choose an in-between size as we did in the ported example. In this case, the three b2 curves use a 2 V_{of} = 1.45 ft³ box and the three b5 curves use a 5 V_{of} = 3.64 ft³ box.

For bandpass systems, a reasonable goal is bass response from 40 to 100 Hz. (This means that the –3-dB frequencies should be below 40 Hz and above 100 Hz, respectively. Note that the –3-dB points are properly referenced to the center frequency level of each curve; thus they are at 0 dB for the +3-dB curves, –3 dB for the 0-dB curves, and –6 dB for the –3-dB curves.)

Looking at the curves closely, only the b2(0) and b2(–3) designs meet the 40- to 100-Hz goal. The b2(0) design is the clear winner over the b2(–3) design because of its 3-dB-higher sensitivity and flatter passband response.

Although all three of the b5 curves easily meet the 40-Hz goal, they fall short at the 100-Hz end. If your main speakers have no problem providing output down to 80 Hz and you have space for a bigger box, then you could use any of the b5 designs and get deeper bass from the subwoofer.

To determine the box frequency, we need to refer to the proper table. We chose the b2(0) design, so we will use Table 4-19. The 0-dB columns for Q_{ts} = 0.3 give us:

$$\text{ported fraction of volume} = 0.478$$

$$f_b/f_{ob} = 1.024$$

This lets us calculate the ported volume and box frequency as follows:

$$\text{ported volume} = 0.478 \ (1.45 \ ft^3) = 0.69 \ ft^3$$

$$f_b = (f_b/f_{ob}) \ f_{ob} = 1.024 \ (69.6 \ Hz) = 71.3 \ Hz$$

These two numbers will be used for box and port construction. The "Box Construction" section later in the chapter explains how.

The Future of Box Design for Car Stereo

Today, the procedure for car subwoofer box design is identical to that used for home speakers.

The problem with this approach is that the interior volume of cars and trucks can be small enough to affect the bass loading of a subwoofer, particularly when all the windows are rolled up. The good news is that you will usually end up with deeper bass than you expected in your car, especially for sealed boxes. The bad news is that you may have built a larger box than you needed, or tuned your port suboptimally.

Eventually, software will become available that lets you factor in the particulars of your vehicle. Some manufacturers of custom truck boxes claim to consider these effects in their products today.

Multiple Drivers in One Box

Using two or more identical woofers in a single subwoofer cabinet can provide a number of advantages over single subwoofer boxes. Multiple driver formats can be used with sealed, ported, or bandpass designs. The five basic configurations are shown in Table 4-23.

Standard. The standard configuration can be thought of as two identical subwoofer systems joined together with the separating wall removed. This doubles the size of your enclosure and the power handling as well. Because the cone area is doubled, the acoustic sensitivity is 3 dB higher than that of a single driver.

This approach can be used with sealed, ported, or bandpass designs. Follow the same box design procedure as you would for a single driver, but assume that you are using a driver having a V_{as} value twice that of an actual single driver. This method will end up giving you a box that is twice the original size, and any port calculations will come out right.

TABLE 4-23

Multiple Driver
Configurations

Configuration	Diagram	Comments
Standard	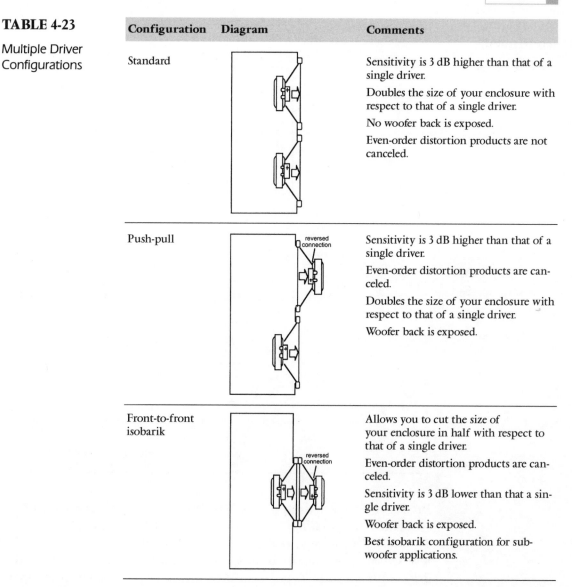	Sensitivity is 3 dB higher than that of a single driver. Doubles the size of your enclosure with respect to that of a single driver. No woofer back is exposed. Even-order distortion products are not canceled.
Push-pull		Sensitivity is 3 dB higher than that of a single driver. Even-order distortion products are canceled. Doubles the size of your enclosure with respect to that of a single driver. Woofer back is exposed.
Front-to-front isobarik		Allows you to cut the size of your enclosure in half with respect to that of a single driver. Even-order distortion products are canceled. Sensitivity is 3 dB lower than that a single driver. Woofer back is exposed. Best isobarik configuration for subwoofer applications.

**TABLE 4-23
(continued)**

Multiple Driver
Configurations

Configuration	Diagram	Comments
Back-to-back isobarik		Allows you to cut the size of your enclosure in half with respect to that of a single driver. Even-order distortion products are canceled. No woofer back is exposed. Sensitivity is 3 dB lower than that a single driver. Construction is more difficult. Heat from both drivers is trapped, reducing power handling. Not recommended.
Piggyback isobarik		Allows you to cut the size of your enclosure in half with respect to that of a single driver. No woofer back is exposed. Sensitivity is 3 dB lower than that a single driver. Even-order distortion products are not canceled. Construction is more difficult. Heat from front driver is trapped, reducing power handling. Not recommended.

Push-Pull. The push-pull configuration is a simple variation of the standard configuration, and shares the characteristics explained above. The only difference is that the direction of one of the two woofers is reversed (and the polarity of its speaker connections is reversed as well, to drive both speakers in phase acoustically). This type of setup will cause even-order nonlinearities to cancel, and will result in substantially lower distortion.

A disadvantage of this approach is that you must look at the back of one of your drivers. Some driver manufacturers enhance the appearance of their subwoofer backs just for this purpose. This is only a cosmetic concern—the sound coming from the back of a subwoofer is just as good as that from the front at subwoofer frequencies.

Front-to-Front Isobarik. Isobarik (which means constant pressure) designs were first described by Harry Olson in the early 1950s. The primary benefit of any isobarik configuration is that it allows you to cut the size of your enclosure in half with respect to that of a single driver. This can be valuable if you are trying to squeeze a subwoofer box into a tight space. It comes at the price of having to buy a second driver and power it. The acoustic sensitivity of an isobarik pair is also 3 dB lower than that of a single driver (due to the doubling of the effective cone mass). This is 6 dB lower than the standard or push-pull configurations for the same number of drivers and amps.

Of all the isobarik configurations, front-to-front is the most useful for subwoofer applications. (The back of a driver is as good as the front for deep bass.) In addition to cutting the enclosure size in half, the front-to-front setup also provides the even-order distortion canceling characteristic of the push-pull configuration.

Isobarik loading can be used with sealed, ported, or bandpass designs, although in practice bandpass isobarik designs can be particularly difficult to get right. Follow the same box design procedure as you would for a single driver, but assume that you are using a driver having a V_{as} value half that of an actual single driver. This method will end up giving you a box that is half the original size, and any port calculations will come out right.

When mounting the drivers, a spacer must normally be used between the two drivers to prevent the two surrounds from touching each other. You can use a ring of medium-density fiberboard (MDF) with appropriately spaced holes to pass the mounting bolts/screws through. Lay the bottom driver in the box after wiring it up (this driver should have its leads wired normally). Lay the MDF ring on top of this driver. Invert the second driver over the first, line up the mounting holes, and screw the whole assembly to the enclosure. The outer driver should have its leads wired in a reversed fashion. This will assure that both drivers are moving in the same direction when a voltage is applied. If you hook everything up and get no bass from your subwoofer, chances are that you've got a driver's polarity reversed.

Back-to-Back Isobarik. The back-to-back isobarik configuration requires a tunnel subenclosure to tightly couple the two drivers. The chief advantage of this approach over the front-to-front isobaric is cosmetic—you get to look at the front of a woofer rather than the back. As with the push-pull and front-to-front isobarik configurations, even-order distortion products are inherently canceled.

Because of the relatively large mass of air coupling the two drivers, accurate frequency response predictions become more difficult. The tunnel makes construction more challenging as well. A further disadvantage of this approach is that the heat generated by both magnet structures is essentially trapped in the subenclosure. This will greatly reduce the thermal power handling of both drivers. For these reasons, this configuration is not recommended.

Piggyback Isobarik. The piggyback isobarik configuration also requires a tunnel subenclosure to tightly couple the two drivers, although it is smaller than that required for the back-to-back isobarik. As with the back-to-back isobarik approach, you get to look at the front of a woofer rather than the back. Unlike with the other isobarik configurations, even-order distortion products are not canceled.

The piggyback isobarik setup shares a number of disadvantages with the back-to-back isobarik. Because of the relatively large mass of air coupling the two drivers, accurate frequency response predictions become more difficult. The tunnel makes construction more challenging. The heat generated by one of the magnet structures is essentially trapped in the subenclosure. This will greatly reduce its thermal power handling and will cause a performance imbalance between the two drivers as one heats up more than the other. For these reasons, this configuration is not recommended.

Box Construction

Whether you buy a pre-made box or build your own, this section contains useful information on mounting drivers, damping enclosures, and constructing ports.

Box Materials and Shapes

A good speaker box should be both rigid and nonresonant. Choosing the right box material is an important step in ensuring this.

Subwoofer boxes should be constructed of ¾-inch-thick particle board or medium-density fiberboard (MDF). Both of these materials are rigid, nonresonant, uniform, easy to work with, and, best of all, cheap. MDF is approximately 35 percent stronger than particle board, but is more expensive and harder to find. Other materials such as hardwood, plywood, or oriented strand board (OSB) may be used, but none are as good as particle board or MDF.

Rectangular boxes are the easiest to build, but to take advantage of an odd-sized space, you may want to use a wedge or other shape. For subwoofers, the shape and relative dimensions of the box do not affect performance. You can build a perfect cube, a tall, thin box, or a pyramid—it doesn't matter, so long as the volume is the same. (This is not true for full-range loudspeakers, where the shape influences mid-bass and higher-frequency response.)

If you want a cylindrical enclosure, use large cardboard tubing called Sonotube® (manufactured by Sunoco for forming concrete pillars). It's available at construction supply outlets. The cylindrical shape provides superior rigidity even though its wall thickness is only ¼ to ½ inch. You can also use large-diameter PVC pipe, available in diameters as large as 12 inches from pipe suppliers. Cut circles of particle board or MDF to fit inside the ends and fasten with Liquid Nails or similar construction adhesive.

> **Important:** When calculating the volume of an enclosure, be sure to use the internal (not external) dimensions. Subtract the volume of items that reduce the internal volume, such as bracing and the back of the woofer itself. For ported boxes, the volume of the port should be deducted too. These volume adjustments normally amount to about a 5 percent reduction in the internal box volume.

Assembly and Bracing

The most popular method of box construction today is to build a six-sided box without removable access panels. This provides maximum strength and makes the box more airtight. Access to the inside of the box is provided through the speaker cutout. The speaker must be mounted from the outside with this method.

Bracing refers to fastening strips of wood (usually 2 × 2 lumber) along internal seams, across panels, or between panels. Bracing strengthens the box and improves its performance by reducing the sound radiated by the box itself. Corner bracing is important where the precision of cuts for the box pieces was limited due to the use of a hand saw, jigsaw, or circular saw. Bracing across panels is beneficial for large enclosures (greater than 1 ft³).

Because of the speaker cutout, the front panel is usually the weakest. It should either be braced or constructed from a double thickness of particle board. If you use the double thickness approach, you may want to precut the speaker cutout in each thickness before gluing them together, depending on the capabilities of your jigsaw.

All box joints and bracing should be glued generously and then nailed or preferably screwed together. Use a wood glue such as Elmer's® Carpenter's Wood glue. Two-inch drywall screws work well for this application. If you don't have an electric drill and screwdriver bit for driving screws, you can use 6d (2-inch) finishing nails.

Bandpass Box Construction

Bandpass designs present some unique challenges when it comes to box construction.

In other types of designs, the woofer is attached from the outside so the box can be constructed without any removable access panels. This is not the case for bandpass boxes. Furthermore, a bandpass system requires sealed and ported chambers. The normal construction technique for a bandpass box is to build the outer cabinet, then add an inner partition to create the two chambers. The removable access panel can be on the sealed side or the ported side (Fig. 4-12). If you want to see your woofer in action, you can use ³⁄₈- or ¹⁄₂-inch-thick Plexiglas for the access panel. It should be sealed with foam gasket material or nonhardening caulk and then secured with a generous number of screws.

For the most part, woofer direction isn't critical. The front of a woofer provides the best high-frequency response. For full-range systems, good high-frequency response is an asset, but for bandpass systems it's a liability. For this reason, it's better to have the back of the woofer in the ported chamber. (If you use a crossover with a steep slope, this factor shouldn't matter.) A second benefit of this orientation is that it provides better cooling for the motor structure of the driver. This points to putting the access panel on the sealed chamber.

Figure 4-12
Removable access panel construction.

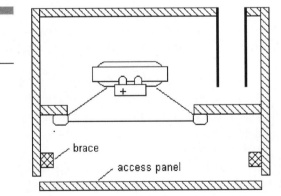

Constructing Ports

Ports can be constructed of PVC pipe (the large-diameter plastic pipe plumbers use). You can also buy prefab speaker ports that you trim to the desired length (Fig. 4-13). There is also an adjustable type that has a sliding outer tube (Fig. 4-14). The advantage of prefab ports is their attractive appearance—they have a contoured mounting flange and are black. The disadvantage is that the length you need in the diameter you want may not be readily available.

Here's the equation to calculate the port length you need:

$$L = \frac{2117\ D^2}{f_b^2 V_b} - 0.732D$$

where port length L and internal diameter D are in inches and box volume (V_b) is in ft^3.

Figure 4-13
Fixed prefab ports.
(*Courtesy of MCM
Electronics.*)

Figure 4-14
Adjustable prefab
port. (*Courtesy of MCM
Electronics.*)

All you need to do is plug in the internal diameter of the port you want to use, the box frequency (calculated in the "Box Design" section), and the box volume. (For a bandpass enclosure, the box volume refers to the volume of the ported section only.)

What port diameter should you use? PVC pipe is available in a number of diameters, including 1.5, 2, 3, 4, and 6 inches (inner diameter). Prefab ports cover a similar range, but often have oddball diameters. Small-diameter ports limit the amount of air passing through them and thus have reduced effectiveness. They are also more prone to audible vent noise. In general, the bigger the better, but as a practical matter, try to use a port diameter that is one-third to one-half the woofer diameter. This is especially important for bandpass boxes because all of the output is from the port.

Large port diameters usually generate extremely long ports, so you may need to limit yourself to a smaller practical diameter. PVC elbows let you create a PVC port that is longer than the box itself by having one or more 90 degree bends. Prefab ports don't offer this flexibility. Use a hacksaw to cut PVC pipe to the desired length and use PVC cement to attach elbows.

Ports may be located on any side of the enclosure. Ports should be placed a minimum of 4 to 6 inches from the woofer to prevent interaction between the cone and the vent. The back of the port should be a minimum of 3 inches from the opposing wall to prevent a reduction of airflow (Fig. 4-15).

To install the port, hold the small end of the port tube against the box where you want it to go and trace around it with a pencil. Use a jigsaw to cut just inside the pencil mark to provide a snug fit. You will then need to file or sand the hole to make it exact. Use Liquid Nails or a similar construction adhesive to glue the tube to the hole.

Tip: Allow adhesives to cure before installing your woofer, since the fumes released by some sealants during curing may have an appetite for foam surrounds.

Port Design Example

Now we'll finish the port design for the bandpass box used as an example in the "Box Design" section. For the Madisound 12204DVC 12-inch

Figure 4-15
Minimum port spacing.

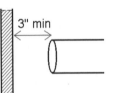

3" min

subwoofer, we previously calculated the ported volume and box frequency:

$$\text{ported volume} = 0.478 \ (1.45 \text{ ft}^3) = 0.69 \text{ ft}^3$$

$$f_b = (f_b/f_{ob}) \ f_{ob} = 1.024 \ (69.6 \text{ Hz}) = 71.3 \text{ Hz}$$

Plugging these in to the port length equation, we get:

$$L = \frac{2117 \ D^2}{f_b^2 \ V_b} - 0.732D$$

$$= \frac{2117 \ D^2}{(71.3)^2 \ (0.69)} - 0.732D$$

$$= 0.604D^2 - 0.732D$$

$$= 6.74 \text{ inches for 4-inch PVC}$$

$$= 17.35 \text{ inches for 6-inch PVC}$$

Since our woofer has a diameter of 12 inches, we would like to have a port diameter of one-third to one-half that, or 4 to 6 inches. For this system, a 4-inch diameter PVC port would need to be 6.74 inches long, while a 6-inch diameter PVC port would need to be 17.35 inches long. Considering that the ported section of the box is only 0.69 ft³, the 4-inch diameter port seems like the right choice.

Multiple Ports

Multiple ports can be used when the look of several smaller ports is preferred to that of a single large port.

There are two widely used methods for calculating the length of multiple ports for a single chamber. Only one method is correct—but, unfortunately, it is the least commonly used.

The incorrect method says that the effective port diameter is equal to the square root of the sum of the squares of the actual port diameters. This is equivalent to saying that we can simply add the cross-sectional areas of multiple ports together and treat them as a single port with the combined cross-sectional area. This sounds reasonable, but it can lead to errors, as we'll see below.

The correct way to figure out how long each port should be follows this simple two-step procedure:

1. Divide the chamber volume by the number of ports you wish to use for that one chamber.

2. Use that as your V_b (box volume) in the port formula, and calculate how long each port should be.

Let's take the previous example using the Madisound 12204DVC 12-inch subwoofer. We'll replace the 4-inch diameter port with four 2-inch diameter ports and see what the two methods give us.

Incorrect Method

According to this method, the effective port diameter of four 2-inch ports is equal to that of a 4-inch port, since both have the same total cross-sectional area. Plugging this into the port formula yields a port length of 6.74 inches—the same result we calculated previously.

Correct Method

We want to use four ports, so we divide 0.69 cubic feet by 4. V_b now becomes 0.173 ft³. Plugging this into the equation using a port diameter of 2 inches gives us a port length of 8.16 inches.

The incorrect method neglects the fact that the end effect (the 0.732D term of the port equation) changes with port diameter.

Speaker Terminals

Speaker terminals should be used rather than running wire through a hole drilled in your speaker box. This reduces air leakage and lets you easily disconnect your speakers.

The most popular terminal types are the spring-loaded pushbutton type (Fig. 4-16) and binding/banana posts (Fig. 4-17).

Figure 4-16
Spring-loaded push-button terminal.
(*Courtesy of MCM Electronics.*)

Figure 4-17
Binding/banana posts.
(*Courtesy of MCM Electronics.*)

The spring-loaded pushbutton types are cheaper and easier to use, but provide less reliable connections and break fairly easily (Table 4-24). Both types accept standard speaker wire, but binding/banana posts also accept spade lugs and banana plugs (Fig. 4-18). These can be soldered onto the ends of speaker wires for a superior connection.

It's best to recess mount terminals. This lets you mount the terminals on the bottom of your box or on a side of it that rests against a wall. It also reduces the chance of breaking the terminals or accidentally releasing them. Recessed mounting cups with built-in terminals are commonly available, but their thin plastic construction allows sound to radiate through them from inside the enclosure. The best approach is to cut a 3-inch round or square hole where you want the terminal, then glue and screw a piece of ¾-inch particle board behind it (inside the box). Mount your terminal to the recessed particle board panel.

Make It Airtight

It's important to make your speaker enclosure as airtight as possible. This applies to sealed as well as ported and bandpass enclosures. Air leaks degrade the box performance and can cause whistles and other noises.

Caulk all internal box seams, the speaker terminals, and the port where it enters the box. Use a siliconized exterior latex caulk for long life and easy cleanup.

Air leaks frequently occur where the woofer mounts to the box. Ironically, woofers have a gasket on their front face, but most woofers are mounted from the outside of the box so this gasket is on the wrong side. Install a foam gasket or use a nonhardening caulk between the box and the woofer to prevent air leakage. DO NOT use regular caulk or silicon rubber sealant to glue the woofer to the box. You will destroy the woofer when you later try to remove it for access inside the box. (Trust me, you WILL want to get inside the box again—you may need to replace a broken speaker terminal or a foreign object may have mysteriously found its way through the port inside your subwoofer box.)

Tip: Do not tin the ends of your speaker wires with solder unless you actually solder them to spade lugs, banana plugs, or the like. Solder is a relatively soft alloy and will readily cold flow under pressure, causing compression connections (such as binding posts or unsoldered crimp connectors) to eventually loosen.

Tip: Allow caulk to cure before installing your woofer since the fumes released by some sealants during curing may have an appetite for foam surrounds.

TABLE 4-24

Types of Speaker Terminals

Speaker Terminals	Comments
Spring-loaded pushbutton	Cheaper and easier to use
Binding/banana posts	Provide better connection, less prone to breakage

Figure 4-18
Banana plugs.
(*Courtesy of MCM
Electronics.*)

In addition to using a gasket or nonhardening caulk to prevent air leak-age, the woofer should be mounted tightly to the box. Wood screws are acceptable, but their holes easily strip out if you aren't careful. A better alternative is to use tee nuts and machine screws. Tee nuts (Fig. 4-19) are inserted into the back of a drilled screw hole and then tapped in with a hammer so their prongs hold them in place. For a high-tech look, use socket-head-cap (Fig. 4-20) or button-head-cap (Fig. 4-21) machine screws.

For isobarik boxes with front-to-front woofers, both woofers are nor-mally mounted outside the box, as shown in Fig. 4-22. When mounting the drivers, a spacer must normally be used between them to prevent the two surrounds from touching each other. You can use a ring of medium-density fiberboard (MDF) with appropriately spaced holes to pass the

Figure 4-19
Tee nuts.

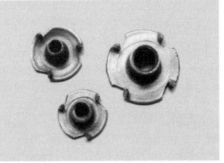

Figure 4-20
Socket-head-cap
machine screws.

Figure 4-21
Button-head-cap
machine screws.

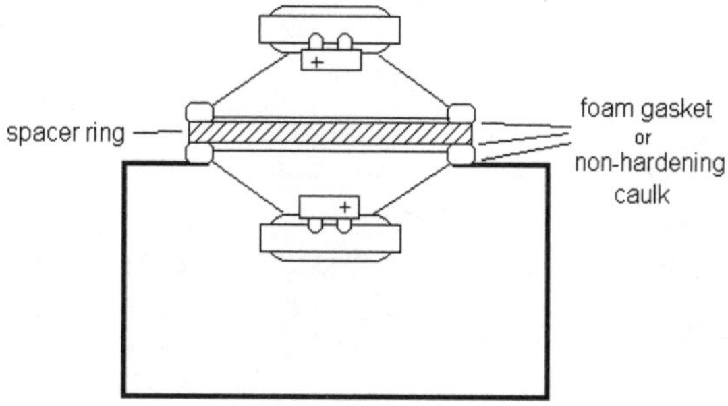

Figure 4-22
Isobarik gasketing.

spacer ring

foam gasket
or
non-hardening
caulk

mounting bolts/screws through. Use a foam gasket or nonhardening caulk where one woofer meets the box and where each woofer contacts the spacer ring. Tee nuts and machine screws are highly recommended for mounting.

Enclosure Damping

Enclosure damping refers to using fiberglass, polyester, or dacron batting or similar materials to line or stuff a speaker enclosure.

Enclosure damping is traditionally used for two reasons. First, it reduces the effects of internal reflections within a speaker enclosure, which can cause resonant peaks and dips in the frequency response at mid-bass and higher frequencies. One to two inches of fiberglass are normally used to line the internal walls of an enclosure for this. This is important for full-range systems, but because subwoofers are used only for very low frequencies, lining them with damping material provides no benefit.

The second use of enclosure damping is to increase the effective size of a sealed box. Fully stuffing a sealed box can increase its effective size by 15 to 25 percent. This can be a useful technique where space is extremely tight, but otherwise isn't worth the trouble. Care must be taken to keep damping material out of and away from the rear basket of the speaker itself. If you plan on stuffing, use only materials that do not shed short fibers. These can find their way into the voice coil gap and cause rubbing and distortion in the speaker. Enclosure stuffing should *not* be used for ported or bandpass boxes.

Finishing Touches

Many products are available to give your handcrafted subwoofer box a professional appearance and protect it from damage. These include custom carpet (Fig. 4-23), corner protectors (Fig. 4-24), handles (Fig. 4-25), and grilles (Fig. 4-26).

Speaker carpet is available in colors to match every automotive interior. It provides a more durable finish than paint and conceals minor imperfections in the cabinet. Carpet is normally applied to all sides of a speaker box using spray adhesive, hot glue, or a staple gun.

Figure 4-23
Custom carpet.
(*Courtesy of MCM Electronics.*)

Figure 4-24
Corner protector.
(*Courtesy of MCM Electronics.*)

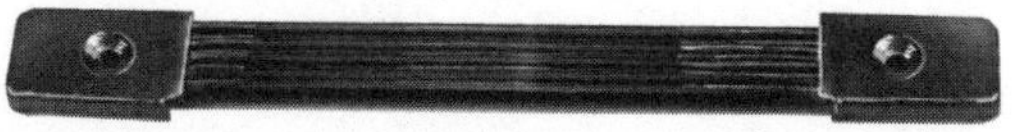

Figure 4-25
Handle. (*Courtesy of MCM Electronics.*)

Figure 4-26
Grilles. (*Courtesy of MCM Electronics.*)

Plastic corner protectors reduce damage to the box and other objects if you bang a corner against something. They also make corners look better—it's difficult to make a corner carpet seam look good otherwise. Plastic or metal speaker grilles protect your expensive woofers from damage from objects in your trunk, dropped screwdrivers, and pets.

Installation

Refer to Chap. 6 for how and where to install the amp for your subwoofer system. In general, the subwoofer crossover should be located close to the subwoofer amp to prevent ignition noise pickup.

Installing a subwoofer system also involves installing the subwoofer box and adjusting the system. This section tells you how.

Bass Blocking Crossovers

Adding speaker-level bass blocking crossovers in line with each of the main (especially the front) speakers is recommended when it's impossible to use a preamp-level high-pass crossover for those channels. This is the case with most head unit speaker outputs. This will reduce the burden of deep bass on the head unit and main speakers, allowing higher volumes with less distortion. The subwoofer will pick up the slack.

For systems where head unit preamp outputs are used for the subwoofer, simply install a bass blocking crossover in series with each speaker (Fig. 4-27).

When using head unit speaker outputs for the sub, be sure to connect the subwoofer crossover inputs to the head unit side of the bass blocking crossovers to prevent blocking bass to the subwoofer (Fig. 4-28).

Bass blocking crossovers should be second-order high-pass filters. You can buy them pre-made or build your own (Fig. 4-29). In either case, you'll need to know the nominal speaker impedance (usually 4 ohms) and decide on the cutoff frequency you want. Use something close to 100 Hz for most applications. Choose a higher frequency when the main speakers are weak in bass or unusually small.

If you want to build your own, refer to the box below. The formulas allow you to calculate capacitance and inductance values for any speaker impedance (R) and cutoff frequency (F) you want. The table in Fig. 4-29 shows three possibilities, based on standard capacitor values and 4-ohm speakers. Capacitors should be of the nonpolarized (bipolar) type.

Tip: If you can't find a capacitor with the value you want, you can wire capacitors in parallel (Fig. 4-30). The capacitance of parallel-wired capacitors is equal to the sum of the individual values.

Bass Blockers

Bass blocking capacitors (or bass blockers) are commonly used as a low-cost method of keeping deep bass from getting to front speakers. Ironically, bass blockers may actually *increase* the amount of deep bass getting to the speakers.

How can this be? If the impedance of a speaker looked like a 4-ohm resistor at all frequencies, then bass blocking capacitors would work as advertised—they would produce a 6-dB/octave high-pass filter with the specified cutoff frequency. In reality, the impedance of a speaker looks nothing like a 4-ohm resistor near the resonant frequency of the loaded driver. The capacitance of the bass blocker interacts with the inductance of the speaker below resonance to produce a low-frequency response peak that can easily be 3 dB higher than with no bass blocker at all!

The bottom line on bass blockers is: DON'T USE THEM! If you want to block deep bass to a speaker and must use a speaker-level filter, use a second-order crossover for the job. Second-order crossovers are relatively immune to the problem just described because they put an inductor in parallel with the speaker. This inductor effectively swamps out the high impedance of the speaker near resonance. Second-order crossovers also provide a superior 12-dB/octave slope.

Figure 4-27
Bass blocking
crossover locations.

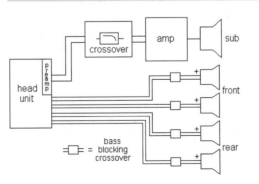

Figure 4-28

Don't block bass to
the subwoofer!

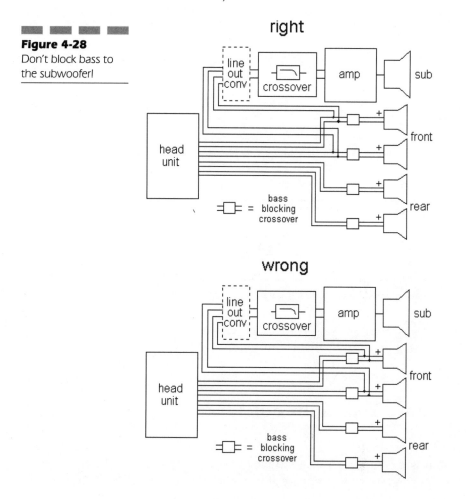

Figure 4-28

Don't block bass to
the subwoofer!

Trunk Sonic Isolation Check

If you plan to use your subwoofer box in the trunk, you should perform
the following check.

Have a friend with a deep voice talk into your open trunk from the
outside. You should be listening inside the passenger compartment with
the windows rolled up. If you can hear your friend loud and clear, then
you'll have no problem using a subwoofer in the trunk. If you can barely
hear your friend, then you may want to provide a better path for the
sound to get from the trunk to the passenger compartment.

One method is to cut a hole in the rear deck and conceal it with a
speaker grille. Another is to cut a hole in the panel behind the back seat.

Figure 4-29
Build your own bass
blocking crossover.

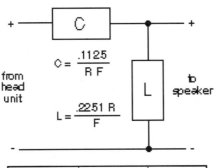

Bass Blocking Crossover

$$C = \frac{.1125}{R\ F}$$

from
head
unit

to
speaker

$$L = \frac{.2251\ R}{F}$$

Freq (Hz)	C (µF)	L (mH)
188	150	4.8
141	200	6.4
94	300	9.6

Table values shown for a 4-ohm speaker

Figure 4-30
Parallel-wired
capacitors.

Secure Your Subwoofer

Whether you've got a tube enclosure that wants to roll around or a box that wants to slide, it's a good idea to secure your subwoofer.

Nylon tie-down straps are available for this purpose and include vehicle mounting hardware. They do a great job of securing an enclosure of any size and shape without screwing into it. Velcro® hook and loop fasteners (Fig. 4-31) work well for keeping boxes from sliding around. When the hook side of the strips is stapled to the subwoofer box, the vehicle's carpeting can act as a suitable loop side. Either of these approaches lets you easily remove the subwoofer when you temporarily need some extra room.

A final option is to use flat or 90° mounting brackets, available at any hardware store (Fig. 4-32). These are screwed or bolted to the subwoofer and vehicle.

Turning It On the First Time

IMPORTANT: Before turning the system on for the first time, turn the volume controls of the subwoofer amp and the head unit to minimum.

Set the crossover cutoff frequency control to the highest frequency. After turning the system on, slightly increase the head unit volume to check that sound is coming from each of the main speakers. If it is not, check for shorted or open connections.

If sound comes from each of the main speakers, increase the subwoofer power amp level control. Listen for bass from the subwoofer. If little or nothing is heard and you are using speaker-level crossover inputs, try setting the head unit's balance control all the way to one side. If this produces bass, one of the subwoofer crossover's inputs has been connected to the system backward. If bass is still not heard, check for +12 volts at the subwoofer crossover power wire and power amp supply and remote turn-on wires. Inspect for shorted, open, or reversed speaker connections.

System Adjustment, or the Secret to Tight Bass

Proper system adjustment is just as important as choosing the right components and hooking them up correctly. In fact, a properly adjusted inexpensive system easily sounds better than a more expensive one not adjusted right.

The secret to achieving tight bass with a system having a subwoofer is to set and keep the head unit's bass control slightly below the central flat response position. The subwoofer is then adjusted to fill in.

Figure 4-31
Velcro. (*Courtesy of MCM Electronics.*)

Figure 4-32
Mounting brackets.

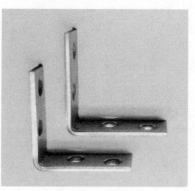

This provides two important benefits. First, it reduces the burden of bass on the main speakers and amps. (The subwoofer will pick up the slack.) This lets you play the stereo louder with less distortion. Second, it provides tighter, less boomy bass. Vehicle interior acoustics are notorious for overemphasizing upper bass frequencies. By turning the bass control down, you reduce this boomy-sounding upper bass. Adjusting the subwoofer crossover cutoff frequency for optimum sound allows the subwoofer to fill in with a deep, solid-sounding bottom end.

Perform the adjustments in this order for best results. Use your favorite cassette, CD, or radio station. Use several sources to get a good average.

Initial Settings.   Preset the subwoofer crossover to 100 Hz. If there is a bass boost feature, turn it off for now. Adjust the subwoofer amp level control to the point that sounds best, erring on the side of too little bass.

Crossover Polarity Switch.   If there is a subwoofer polarity switch, set it to whichever position gives you the most bass. The effect is usually small.

Crossover Bass Boost.   If you plan to use the bass boost feature, turn it on now. In general, it will help extend the system bass response and provide tighter-sounding bass. If you later find that your subwoofer sometimes distorts, you should turn the bass boost off.

Crossover Cutoff Frequency and Power Amp Level.   Now fine-tune the cutoff frequency of the subwoofer crossover and the level control of the subwoofer power amplifier to provide the most natural-sounding bass. The bass control of the head unit should be slightly below the flat position as you do this.

You will find that you need more bass when you are driving to overcome the masking effects of road noise. (See the box on road noise masking in Chap. 7.) Try to find a level that is a good compromise for all conditions. After you're satisfied with the settings, you can use the bass control of the head unit to compensate for individual cassettes, CDs, and radio stations.

Bass Transducers

Bass transducers are a relatively new development in car audio. Working on the principle that low bass is mostly felt and not heard, bass transduc-

Figure 4-33
Aura Bass Shaker.
(Courtesy of Aura.)

ers directly vibrate the chassis of your vehicle. Two examples of bass transducers are Aura Bass Shakers (Fig. 4-33) and Alpine Bass Engines.

Bass transducers are most effective when bolted to the floor beneath the seats. This provides good coupling to the entire vehicle. (Mounting them directly to the seats may seem like a better idea, but you can feel where the source of the vibration is.)

Bass transducers have a 4-ohm impedance, and are designed to be driven like ordinary subwoofer drivers. Any car stereo amp with the desired power can be used—25 to 50 watts per channel is normally recommended. Although not strictly required, a crossover is advised to prevent any audible vibrations. (An amp with built-in sub crossover is ideal.) As with a subwoofer, the level of the amplifier must be properly set for a realistic effect.

Since bass transducers produce vibration rather than sound, they are best used in conjunction with subwoofers rather than instead of them. What bass transducers can give you is the sensation of extremely powerful bass with zero boominess.

Bass transducers are ideal for open-air vehicles such as convertibles, Jeeps, or Trackers. In these vehicles it is almost impossible to achieve the sensation of solid bass with traditional subwoofers alone because you are radiating into open air. Bass transducers are also well suited for vehicles lacking the space for a large subwoofer, such as two-seater sports cars and small pickup trucks. By using a bass transducer, you can get away with using a smaller subwoofer having a higher cutoff frequency.

Head Unit Projects

Head unit is the industry term for an in-dash radio or receiver. Adding or upgrading a head unit along with a new set of speakers is undoubtedly the most popular car stereo project—not surprising, since the head unit is the heart of any car stereo system. With today's convenient wiring harness adapters and mounting kits, head unit projects are easier than ever.

If you have a premium factory sound system (such as Delco/Bose or Ford/JBL) and want to replace your head unit, the situation is slightly more complicated. See the section on premium factory sound systems later in this chapter for details.

Why Choose an Aftermarket Head Unit?

Many factory head units are lower-power units of about 3 to 5 watts per channel. Even if the factory system is a high-power unit, there are still reasons to go with an aftermarket head unit. One good reason is easier integration with equipment such as disc changers and amplifiers.

Factory speakers should not have a problem handling the output from a high-power head unit. However, since speaker performance is so critical to listening enjoyment, adding quality aftermarket speakers will let you take advantage of the improved sound your new head unit has to offer.

Pick One That Fits

Before shopping for head units, you'll need to determine what size, mounting method, and depth your vehicle can accommodate. First, it's important to understand the basics of how the aftermarket head unit industry supports the wide diversity of mounting applications with just a few standard sizes.

Size and Mounting Method Basics

Factory head units come in many sizes and use a number of different mounting methods. Automobile manufacturers are constantly inventing new sizes and mounting methods, which makes it more challenging to use aftermarket head units. Most aftermarket head units today will directly fit in Euro DIN, ISO-DIN, and wide-face ISO-DIN vehicle applications. With the proper mounting adapter kit, these units will also fit virtually any other late-model vehicle. For older vehicles, universal shaft-style head units must be used to avoid the need for major dashboard modification.

The most important types of head units (size and mounting method) are shown in Table 5-1.

Aftermarket DIN-compatible head units are of the correct size and provide means to support both Euro DIN and ISO-DIN mounting standards. A front adapter plate is included to support the wide-face ISO-DIN-J applications. (Without the adapter, you would have a gap on each side of the head unit in ISO-DIN-J applications.)

Universal shaft-style radios are required for many pre-1984 vehicles. They can also be used in almost any later vehicle with the proper mounting adapter.

Even though DIN head units can be made to fit almost every late-model vehicle with the proper mounting adapter kit, there are cases where this gives the look of a small head unit in a big panel. This is where GM- and Chrysler-style aftermarket head units come in. These are designed to fill the large factory openings in many GM and Chrysler vehicles without a mounting adapter kit.

TABLE 5-1

Common Head Unit Size/Mounting Methods

Name	Comments
Euro DIN	Most common size and mounting method. Flat front design. Fits through a rectangular hole in dash and snaps into place with locking tabs.
ISO-DIN	Same size as Euro DIN, but mounts to a side-support factory frame using screws. Common in Toyota and Nissan.
ISO-DIN-J	Same as ISO-DIN, but with wider face. Common in Nissan and Mitsubishi.
Universal	Shaft style, fits three-hole dash opening. Mounts from behind dash. Required for many vehicles prior to 1984.
GM	Front-mount, oversized-face design. Found in many 1982–1996 GM vehicles.
Chrysler	Front-mount, oversized-face design. Found in most 1981–1997 Chrysler, Plymouth, and Dodge vehicles.
Double DIN	Twice the height of ISO-DIN. Used for combined CD/cassette units. Found in late-model Toyota and Nissan.

What Does DIN Stand For?

DIN stands for *Deutsche Industrie Norm*, which means "German Industry Standard." The DIN head unit standard was created in the seventies and has been widely adopted in Europe. It is used throughout the world for aftermarket head units.

Determining What Your Vehicle Can Accommodate

In addition to head unit size and mounting method, you also need to check mounting depth. Head units have a typical mounting depth requirement that ranges from about 6 to 7 inches. Vehicles can have a mounting depth of as little as 5½ inches to as much as 12 inches. This means that not all head units with the proper face size and mounting method will automatically fit your vehicle. In cases where depth is a problem, a mounting adapter kit with a molded extension may be available.

One way to determine what size, mounting method, and depth your vehicle can accommodate is to inspect and measure the vehicle yourself. This requires familiarity with each of the sizes and mounting methods listed in Table 5-1 and forces you to remove your old head unit before you buy a new one. An easier alternative is to consult with a car stereo shop or use a head unit application guide. The Crutchfield catalog (Fig. 5-1) contains complete information for most vehicles as well as for all the head units Crutchfield carries.

Figure 5-1
Crutchfield catalog.
(*Courtesy of Crutchfield.*)

Cassette, CD, or MiniDisc?

Whether to choose cassette or CD is a matter of personal preference and cost. Another option is the recordable Sony MiniDisc. It offers a digital alternative to cassettes with the convenience of CDs—you can find tracks instantly and you never need to rewind. When it comes to cost, cassette is the least expensive, followed by CD. MiniDisc is the most expensive, but prices are dropping.

MiniDisc Technology

The MiniDisc format combines many of the benefits of both CDs and cassettes. Like CDs, MiniDiscs provide random access, letting you find tracks instantly. Like cassettes, MiniDiscs let you record (and rerecord) your choice of music. You can also buy prerecorded MiniDiscs of many popular titles.

Let's take a look at the unique aspects of the MiniDisc Format.

Small Disc Cartridge

MiniDiscs use a 2.5-inch diameter disc housed in a rigid protective cartridge, much like a 3.5-inch computer disk, but smaller (Fig. 5-2). About six MiniDiscs can fit in the same space as a cassette case. Since the actual disc is never touched, there are no problems with scratches or dirt as with CDs. Although MiniDiscs come in a jewel case, the protective cartridge allows the disc to be transported without the case, saving as much space as possible.

Figure 5-2
MiniDisc. (*Courtesy of Sony Electronics Inc.*)

Recordable and Playback-Only MiniDiscs

There are two types of MiniDiscs available: recordable and prerecorded playback-only.

Playback-only MiniDiscs are designed for prerecorded music and are sold by record companies. Manufactured using the same process as CDs, playback-only MiniDiscs feature full-cover artwork. Since they cannot be recorded, it is impossible to inadvertently record over them.

Recordable MiniDiscs are magneto-optical discs. Only through the combination of laser light and a magnetic field can a recordable MiniDisc be recorded (or rerecorded). This means that a recordable MiniDisc can be rerecorded an almost unlimited number of times, yet recorded information is virtually immune to erasure or degradation due to magnetic fields.

Data Compression

Because of the small size of the MiniDisc, data compression is required to fit 74 minutes of music (the same amount as a CD) on the disc. Without this compression, play time would be limited to about 15 minutes.

Music information is compressed using Sony's *Adaptive Transform Acoustic Coding* (ATRAC) system. This system was designed specifically for high-fidelity audio using the latest in digital data compression technology and results in virtually no loss in sound quality.

If you want both CD and cassette, you have two options. You can put a cassette receiver in the dash and add a CD changer in the trunk or elsewhere. The other option is to install a combined CD/cassette head unit (Fig. 5-3). These require a double-height (double DIN) dash opening, so they only fit in some vehicles.

Figure 5-3
Clarion ADX5355 combined CD/ cassette head unit. (*Courtesy of Clarion.*)

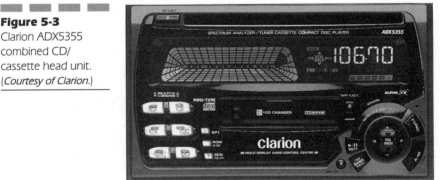

Performance and Features

It's easy to get confused over all the features and performance aspects of head units. This section explains what each feature is all about. For convenience, features are categorized as general features, radio features, cassette features, and CD features.

General Features

The features in Table 5-2 are important to consider in any head unit, regardless of how often you listen to the radio or whether you choose cassette, CD, or MiniDisc in the dash.

TABLE 5-2

Head Unit General Features

Performance/ Feature	Comments
Theft deterrents	Many deterrents are available: fully and partially detachable faces, concealable faces, card key and security code systems.
CD changer controls	Key feature if you want to add a CD changer now or in the future.
Cell phone mute	Automatically mutes the audio when a cell phone is in use.
Front image enhancer	Provides a more authentic sound stage.
Bass and treble controls	A must for quick and effective frequency compensation.
Loudness	Boosts bass for low-volume listening.
Preamp outputs	One, two, or three sets of RCA jacks on the back of the head unit to connect an external amp.
Preamp output voltage	Higher voltage makes it easier to get noise-free performance. Typical range is 500 mV to 4 volts.
Balanced line driver outputs	High performance, but expensive alternative to preamp outputs.
Remote control	Hand-held and mounted types.
RMS power	Look at RMS power rather than peak power for a more accurate indication of performance.
Speaker outputs	Most head units provide four speaker outputs, although some low-end units provide only two, and some high-end units provide none at all.

Theft Deterrents. There are a number of types of theft deterrents available in today's aftermarket head units:

■ Fully detachable face

■ Partially detachable face

■ Concealable face

■ Card key

■ Security code

A fully detachable face lets you take the entire head unit faceplate with you, leaving only a blank panel behind. A partially detachable face lets you take only the essential controls, leaving the bulkier ones behind. The disadvantage of detachable faces is that you need to remove and carry the faceplate with you.

On some Pioneer units, a built-in alarm system is armed each time the faceplate is removed and disarmed when it is reinserted into position. When the alarm is activated, the system emits an ear-piercing tone through the audio system speakers inside the car.

Concealable face head units have a manually operated or motorized fold-down face. Kenwood's MASK™ head units automatically display a blank panel when the ignition is turned off. This approach eliminates the hassle of removing and carrying faceplates; however, it leaves a complete functional head unit for thieves.

Blaupunkt KeyCard™ head units (Fig. 5-4) use a credit-card-size electronic key for protection. The KeyCard must be inserted into its slot for

Figure 5-4
Blaupunkt KeyCard head unit. (*Courtesy of Blaupunkt.*)

the head unit to operate. A KeyCard is easier to carry than a detachable face and is coded with a matching serial number for security. When the KeyCard is not inserted, a blinking card-loading tongue alerts would-be thieves to the system.

User-programmable security codes are commonly used on high-end factory units and a number of aftermarket head units. If power to the unit is ever removed, the security code must be entered to reactivate it. Instead of entering a number, Eclipse head units have you choose one of your CDs as a key CD. This disc must be loaded to reactivate the unit after loss of power.

CD Changer Controls. This is a key feature if you think you might want to add a CD changer now or in the future. Head units with CD changer controls have a special connector on the back for a direct audio connection with the CD changer, for the best possible sound. They provide all the buttons and displays needed to support full operation of the CD changer as well.

In general, you'll need to choose a CD changer from the same manufacturer as the head unit for compatibility. Some controller/changer combinations have special programming features. For example, you can specify exactly which track will play on each CD, a great way to eliminate songs you don't like. Some let you assign each disc a name that will appear on the dash display when that disc is loaded.

There are alternate ways to add a CD changer to a system, but none are as good as using a head unit with CD changer controls. See Chap. 9 for details.

Cell Phone Mute. Automatically mutes the audio when a cell phone is in use. (Requires connection to a mute wire from the vehicle's cellular phone system.)

Front Image Enhancer. This feature, available in some Pioneer head units, provides an enhanced imaging mode for a more authentic sound stage. This is achieved by using the front speakers as primary drivers and the back speakers for deep bass and rear fill. When you want maximum volume or have passengers in the back, you can turn the feature off.

Bass and Treble Controls. Separate bass and treble controls are a must for quick and effective frequency compensation. A single tone control just can't do the job right.

Loudness. This control lets you boost bass frequencies to keep music full-sounding at lower volumes. On some head units it's automatic.

Preamp Outputs. RCA jacks on the back of the head unit that allow you to use a standard patch cable to connect an external amp. (Speaker outputs can be also used to provide signal to some amps, but preamp outputs provide a low-noise, low-distortion interface to any amplifier.)

Many head units have a single set of preamp outputs. Head units with two sets of preamp outputs provide increased flexibility for boosting with amplifiers. Having two sets of preamp outputs means the head unit fader control can remain functional for most configurations. An occasional head unit has three sets of preamp outputs, one of which is intended to be used for a subwoofer amp.

If you plan to boost your head unit with external amps, see Chap. 6 for details on how the number of preamp outputs affects your system configuration options.

Preamp Output Voltage. A higher preamp output voltage makes it easier to get noise-free performance from external amps. The typical range is 500 mV to 4 volts.

Balanced Line Driver Outputs. Balanced line driver outputs go one step beyond high-voltage preamp outputs for providing noise-free performance with external components. They provide differential 8-volt drive, which essentially allows externally induced noise to be subtracted out. Balanced line driver outputs must be used with crossovers and amps having balanced line driver inputs. Balanced signals are normally routed using special XLR connectors and cables. Not for the budget conscious.

Remote Control. There are two categories of remote controls for head units: hand-held and mounted. Hand-held remotes are useful if you want to control the head unit from outside the vehicle or from anywhere in a van or RV. Mounted remotes may be of the type mounted on the steering wheel within reach of your thumb (Fig. 5-5), or they may be mounted on the console or steering column. The idea is to make it easier to control your stereo without taking your hands off the wheel or your eyes off the road.

RMS Power. The amount of continuous power, measured in watts, that an amplifier produces. The higher the number, the louder your music can play without distorting. Manufacturers often show peak power ratings on the faces of their products. Peak power measures the instantaneous power at the peak output voltage of the amplifier. Look at RMS power for a more accurate indication of performance.

Figure 5-5
Blaupunkt Thummer™
remote control.
(*Courtesy of Blaupunkt.*)

High-Power Versus Low-Power Head Units

For head units, there are actually two distinct power categories (Table 5-3). These correspond to two internal amplifier design types: single-ended and internally bridged. These power categories arise because of the 12-volt electrical system of the car.

Single-ended amps are the least expensive type, and are found in many factory and low-cost aftermarket head units. Their power is limited by the fact that one amplifier output terminal is ground (0 volts) and the other terminal has a dynamic range of roughly 12 volts (its maximum swing is limited by the 12-volt power supply). This restricts the power to approximately 5 watts per channel.

Internally bridged amps are found in high-power factory and aftermarket head units. They achieve roughly 4 times the power of the single-ended amps by driving both plus and minus output terminals via separate internal amps. Driving the minus output terminal with an inverted version of the signal at the plus terminal doubles the effective output voltage to the speaker. Doubling the speaker voltage results in 4 times the power.

This also explains why you can't ground one side of the speakers when using high-power head units—the minus side of the speakers is actually being driven by an internal amplifier.

TABLE 5-3

The Differences Between High- and Low-Power Head Units

Head Unit Type	Internal Amps	Power Category
Low-power	Single-ended	5 watts per channel typical
High-power	Internally bridged	18 watts per channel typical

Speaker Outputs. Most head units provide four speaker outputs, although some low-end units provide only two. Some high-end units provide none at all. When comparing models with four speaker outputs, realize that better units use four internal amplifiers—one per channel. Cheaper units use only two internal amplifiers and share them between front and back using a speaker-level fader. Speaker-level faders not only waste amplifier power, they can significantly increase the output impedance of the speaker outputs. This causes boomy-sounding bass. Look for models with four internal amp channels when you need to drive four speakers.

Radio Features

If you spend a lot of time listening to the radio, you'll want to choose a head unit with good FM performance and features. Look for a tuner with low FM sensitivity and high FM selectivity. If you like to channel surf, go for lots of presets and features like preset scan and automatic preset programming. Check out what RDS has to offer too. Table 5-4 gives a summary of the various radio features.

AM/FM Presets. These user-programmable push buttons let you go directly to the station you want. Some head units have as many as 36 presets available.

Diversity Tuning. Uses two antennas and a special tuner to improve reception. It works by automatically selecting the antenna with the best reception. The speed of the selection process is so rapid that you cannot perceive it.

Systems with diversity tuning have excellent multipath rejection. Multipath is that swishing sound you hear when driving around large buildings or the suddenly bad reception you sometimes get when you're stopped at a traffic light. It's caused by a reflected version of the radio signal interfering with the direct signal. Diversity takes advantage of the fact that it is extremely likely that there is good reception at one of the antennas at all times.

ID-Logic. Uses a computer chip programmed with call letters for nearly every radio station in the country. You key in your location and it will scan and load stations by format. It works for both AM and FM stations. This feature is unique to Pioneer head units.

More About Multipath

The nature of a multipath signal is such that the received radio signal can drastically change strength over a distance of only 5 to 10 feet. This is because the typical half-wavelength of an FM signal is about 5 feet, and cancellation often occurs as two signals travel different path distances to reach the same point (hence "multipath"). If there is cancellation at one point, then about 5 to 10 feet away there is usually a useful-strength point.

TABLE 5-4

Head Unit Radio
Features

Performance/ Feature	Comments
AM/FM presets	User-programmable push buttons let you go directly to the station you want.
Diversity tuning	Uses two antennas and a special tuner to improve reception.
ID-Logic	Allows station tuning by format.
FM mono/ stereo switch	Allows you to improve the signal-to-noise ratio during weak signal conditions.
RDS	Enables your head unit to display the station name and other text. Allows station tuning by format. Works only for FM stations transmitting RDS.
Preset scan	Plays a brief sample of what's on each of your preset radio stations.
Station scan	Plays a brief sample of what's on each strong station.
FM mono sensitivity	Tells you how weak an FM station a tuner can pick up.
FM selectivity	Tells you how well a tuner can reject the signal of a nearby interfering radio station.
FM stereo separation	The ability of an FM tuner to recreate the proper stereo image.
Automatic preset programming	Automatically loads your presets with the strongest available signals.

FM Mono/Stereo Switch.   Allows you to revert to mono mode during weak signal conditions to improve the signal-to-noise ratio. On some head units, operation is automatic.

RDS.   Enables your head unit to display the station name, song title, and artist and other text messages that many FM radio stations now include with their broadcast signal. Also allows you to tune radio stations by music format, such as rock or jazz.

More About RDS

RDS stands for *Radio Data System*. This term is commonly used to refer to the decade-old European system as well as the newer, slightly different U.S. system. (The official name of the U.S. system is RBDS, short for *Radio Broadcast Data System*.) Over 700 U.S. FM stations use RDS (as of 1998), most of them in the top 25 markets.

RDS is a system for sending digital information along with the regular FM program. You can receive it using one of the newer tuners that have RDS decoding and an alphanumeric display built in. The RDS data stream can carry lots of useful data, such as:

- Station name (WFBQ or Q95)
- Format (country, rock, jazz, etc.)
- Song title and artist or other text
- Time of day
- Traffic alerts
- Weather alerts

Some stations use RDS just for its basic features, displaying their station name and format on RDS radios. With an RDS radio, a scan looks first for RDS stations. If you further scan by format, then the radio will only stop at RDS stations that identify themselves as broadcasting that format. All of the rest are ignored.

RDS can also update some head units' clocks. With a unit capable of receiving the traffic and weather alert features of RDS, you can be alerted with reports even while listening to a cassette or CD.

RDS radios are not that common—yet. There are only about 1 million in the United States now, but more are coming fast. RDS is an option in GM vehicles and standard in some European imports. Blaupunkt, Clarion (Fig. 5-6), Denon, Kenwood, Panasonic, Philips, Pioneer, and Sony are some of the manufacturers now producing radios with this feature.

Figure 5-6
Clarion DRX9375R
RDS head unit.
(*Courtesy of Clarion.*)

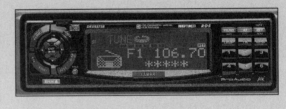

Preset Scan. Touch a button and your receiver automatically plays a brief sample of what's on each of your preset radio stations. Touch the button again when you hear something you want to stay with.

Station Scan. Touch a button and your receiver automatically plays a brief sample of what's on each strong station. Touch the button again when you hear something you want to stay with.

FM Mono Sensitivity. This figure tells you how well a tuner can pick up weak FM radio signals. Lower values are better. Expressed in decibel femtowatts (dBf).

FM Selectivity. Alternate channel selectivity tells you how well a tuner can receive a desired radio station's signal while rejecting the signal of a nearby radio station. It is rated in decibels. Unlike the spec for sensitivity, the higher the number for selectivity the better. A good figure is 75 to 80 dB.

FM Stereo Separation. A measure of the ability of an FM tuner to recreate the proper stereo image. Normally specified at 1 kHz and measured in decibels. Anything 30 dB and above should be considered acceptable.

Automatic Preset Programming. Different manufacturers have different names for this feature. Activate it and the tuner automatically loads your presets with the strongest available signals. This makes finding stations easier when you're traveling.

Cassette Features

If you listen to a lot of tapes, look for a wide frequency response and Dolby noise reduction. Most prerecorded tapes use Dolby B, but if you record your own with Dolby C you may want a deck that offers both B and C. For features, look for auto reverse and auto music search. Choose soft-touch electronic controls if you dislike the feel of mechanical ones. Table 5-5 shows a range of cassette features.

Auto Music Search. Fast-forwards or rewinds to the next song and begins to play automatically. Multitrack auto music search skips forward or backward over multiple songs—hit the fast-forward or rewind button once for each track you want to skip over.

Auto Reverse. Automatically plays the other side of the tape when the end is reached. This feature also lets you switch to the other side of the tape at the push of a button.

Dolby. Noise reduction system to virtually eliminate tape hiss. Head units with this feature decode Dolby-encoded tapes during playback. Dolby B is the most widely used system, but Dolby C offers a greater improvement in signal-to-noise ratio. Dolby is not as important in the relatively noisy automotive listening environment as it is in the home. Note: You have to record with Dolby C to get any benefit from a Dolby C car deck.

TABLE 5-5

Head Unit Cassette
Features

Performance/ Feature	Comments
Auto music search	Fast-forwards or rewinds to the next song and begins to play automatically.
Auto reverse	Automatically plays the other side of the tape when the end is reached.
Dolby	Noise reduction system to virtually eliminate tape hiss.
Tape eq switch	Selects the best equalization response according to the type of tape—normal, high-bias, or metal.
Tape frequency response	The frequency range a tape deck can reproduce.
Power-off release	Disengages the pinch roller from the capstan when you turn off the tape deck. Prevents a flat spot in the pinch roller.
Radio during rewind	Automatically plays the radio when a tape is fast-forwarding or rewinding.
Tape signal-to-noise ratio	A measure of how well a cassette player silences background noise. Almost purely a function of whether the deck has Dolby, and which type.
Soft-touch controls	More luxurious than the traditional spring-loaded mechanical buttons.
Wow and flutter	How stable the tape playback speed is.

Tape EQ Switch. Selects the best equalization response according to the type of tape—normal, high-bias, or metal. In some head units selection is automatic.

Tape Frequency Response. The frequency range a tape deck can faithfully reproduce. Humans can hear sounds as low as 20 Hz and as high as 20 kHz.

Power-Off Release. Disengages the pinch roller from the capstan when you turn off the tape deck or turn off your car without first ejecting the tape. Prevents a flat spot from forming on the pinch roller, as well as a crease in the tape.

Radio During Rewind. Automatically plays the radio when a tape is fast-forwarding or rewinding. On some head units this feature is automatic; on others it's selectable.

Tape Signal-To-Noise Ratio.   A measure of how well a cassette player silences background noise. Ratings are in decibels; higher ratings indicate less noise. This spec is almost purely a function of whether the deck has Dolby noise reduction and which type. Don't be swayed by better specs on decks with Dolby C if you only listen to Dolby B tapes: Chances are the performance with Dolby B tapes is the same as for decks with only Dolby B. In other words, you have to record with Dolby C to get any benefit from a Dolby C car deck.

Soft-Touch Controls.   Soft-touch electronic controls are more luxurious than the traditional spring-loaded mechanical buttons. When you insert or eject a cassette, the deck automatically loads or smoothly unloads the tape. Soft-touch controls also make possible a number of advanced features, such as multi-track auto music search and blank skip.

Wow and Flutter.   This spec indicates how stable the cassette deck's playback speed is. The lower the percentage, the better.

CD Features

A CD player in the dash is the ultimate in sound quality and convenience. Excellent frequency response and signal-to-noise ratio are the norm. Features like intro scan, random play, and track repeat are found on almost all models. If you go off road, a model with electronic shock protection can help reduce skipping. Unless you're a competitor, look to the general features and radio features to help you choose the right CD head unit. Table 5-6 provides an overview of CD features.

Number of Discs.   Single-disc capacity is standard for CD head units. An exception is the JVC KD-GT5 (Fig. 5-7), a head unit with a built-in three-disc CD changer. The KD-GT5 is a DIN-compatible unit, so it will easily fit in most dash openings. This unit has two sets of preamp outputs and requires a separate amplifier.

CD Frequency Response.   The frequency range a CD player can faithfully reproduce. Humans can hear sounds as low as 20 Hz and as high as 20 kHz. Virtually every CD player exceeds this range.

Electronic Shock Protection.   Electronic shock protection is a buffer memory used to store between 1 and 10 seconds of music as a reserve

TABLE 5-6

Head Unit CD
Features

Performance/ Feature	Comments
Number of discs	Single-disc capacity is standard, three-disc capacity is available.
CD frequency response	The frequency range a CD player can reproduce.
Electronic shock protection	A buffer memory used as a reserve against skipping.
Intro scan	Lets you hear the first few seconds of each track. Hit the button again when you hear the song you want.
Random play/shuffle	Mixes up the order of songs for variety during playback.
CD signal-to-noise ratio	A measure of how well a CD player silences background noise.
Zero-bit detect	Mutes the output whenever a series of zeros is detected in the digital bitstream, for complete silence between songs.

against skipping. It can significantly reduce skipping on bumpy roads or off road.

Intro Scan. Lets you hear the first few seconds of each track. Hit the button again when you hear the song you want.

Random Play/Shuffle. Mixes up the order of songs for variety during playback.

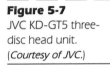

Figure 5-7
JVC KD-GT5 three-disc head unit.
(*Courtesy of JVC.*)

CD Signal-to-Noise Ratio. A measure of how well a CD player silences background noise. Ratings are in decibels; higher ratings indicate less noise. The signal-to-noise ratio of CD players is generally so good that noise picked up by other means determines the system noise level. Unless you're a competitor or plan on using super-high-power amps, don't place too much emphasis on this spec.

Zero-Bit Detect. Mutes the output whenever a series of zeros is detected in the digital bitstream. The result is complete silence between songs. This is important if you're a competitor, but otherwise you'll never hear the difference.

Installation

Once you've determined what size, mounting method, and depth your vehicle can accommodate and what features and performance you want, you're ready to buy your head unit.

The key to making head unit installation go smoothly is having the necessary mechanical and electrical adapters, supplies, and tools on hand. Adapters are explained later in this chapter. For guidance on the supplies and tools you'll need, check out Chap. 2. The best time to buy adapters and supplies is when you buy your head unit.

Installing a head unit is really two projects in one—a mechanical mounting project and an electrical wiring project. Each of these is explained next.

Mechanical Mounting

For many installations, an aftermarket head unit will directly fit the factory mounting system and dash opening. For these, you don't need a special mounting kit or faceplate. For other installations, the key to mechanical mounting is using the right mounting kit.

Mounting Kits. Mounting kits are available to make a DIN or shafted head unit fit in virtually any late-model dash. (For older vehicles, a shafted head unit must be used, and no kit is required.) Different kits are designed to remedy different mounting problems. Some kits are specifically made for shafted installations, some are for DIN only, and others are

convertible (providing a shafted mount with a cutaway opening for DIN). For shallow mounting depths, kits with extended faces can give you additional depth to work with. Some kits have a half-DIN equalizer opening built in.

The Metra *Pocket with a Purpose* GM installation kit (Fig. 5-8) will hold an AM/FM/CD player and a CD jewel case. This solves the problem of where to stash a CD case and is a great way to take advantage of the extra space available with a DIN head unit in a GM slot.

Kits for some vehicles require a lot of exposed plastic. For these, it's important to match the factory appearance as much as possible. The Scosche GM1482 kit for Chevy and GMC full-size trucks (Fig. 5-9) is available in six colors to match the dash.

Some vehicles present special challenges. Starting in 1996, Ford decided to integrate the climate controls and head unit controls into the same dashboard panel in their Taurus and Sable automobiles. The actual head unit is located in the trunk. This made aftermarket head unit installation an extremely difficult proposition in those models. That same year

Figure 5-9
Scosche GM1482 kit. (*Courtesy of Scosche.*)

Figure 5-10
Scosche 1996 Taurus
kit FD1340. (*Courtesy
of Scosche.*)

Scosche responded with the FD1340 Integrated Control Panel kit (Fig. 5-10). This kit made it possible to install an aftermarket head unit in a Taurus or Sable dash while retaining the factory climate controls.

The point of all this is that not all mounting kits are created equal. Ask about your options when shopping.

Tip: Always try to find a way to provide rear support for a head unit. The bumps and vibrations in the automotive environment can reduce the life of an unsecured unit and increase wow and flutter in a cassette deck or skipping in a CD deck. Abuse from potholes and railroad crossings can also put stress on the installation kit and surrounding dash panels.

A backstrap (strip of metal perforated with a series of holes) is often supplied with a head unit for rear support (Fig. 5-11). One end of the strap attaches to a stud on the back of the head unit, the other to an existing bolt or screw under the dash. The multiple holes give you a choice of attachment points. You will normally need to bend or twist the backstrap to reach the point of attachment.

Figure 5-11
Backstrap. (*Courtesy
of MCM Electronics.*)

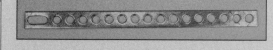

Removing and Installing in a Three-Hole Dash Opening. This method applies to older vehicles (prior to 1984) requiring a shafted head unit. In these vehicles, the two shafts and nosepiece of the head unit protrude from behind through matching holes in the dash.

Before you can install your new head unit, you'll need to remove the old one. Pay close attention to the steps involved—the process for installing the new one will be the same, but in reverse. Pull off the volume and tuning knobs plus any fader or balance control rings behind them, then remove the nuts and washers on the two shafts. You should now be able to work the head unit back through the dash and out. This can be somewhat tricky because of obstructions behind the dash, such as heater controls and ducts.

The new head unit will install the same way, but you may need to make some adjustments to make it fit right. You may need to adjust the spacing between the shafts to match the holes in the dash. Shafted units generally allow you to do this by loosening the innermost shaft nuts, repositioning the shafts to their desired positions, and then retightening. Adjust the next level of shaft nuts so the nosepiece and shafts protrude from the front of the dash by the desired amount when installed with the new faceplate in place. If limited installation depth forces the shafts to protrude too far, plastic spacers can be used on the exposed shafts to improve the appearance of the installation.

Once you've made all the adjustments, push the head unit forward against the dash and put the aftermarket faceplate on, followed by washers and nuts on the shafts. Put the control rings and knobs on the shafts. Don't forget a backstrap, especially if you've got a plastic dash.

Removing and Installing in an Oversize GM or Chrysler Dash Opening. The oversize dash openings found in many GM and Chrysler vehicles give you lots of options because they have extra space to work with. You can choose an adapter kit that lets you install a Euro DIN, ISO-DIN, or shaft-style head unit plus a half-DIN equalizer. You can choose a kit that centers the head unit in a plain panel or one with a storage pocket below the head unit. If you prefer the look of a head unit face that fills the entire dash opening, GM- and Chrysler-style aftermarket head units are available. These are designed to fill the large factory openings in many GM and Chrysler vehicles, and eliminate the need for a mounting adapter kit.

Before you can install your new head unit, you'll need to remove the old one. Normally a factory trim panel will need to be removed first. This might be attached with screws or spring clips, or a combination of the two. Often, screw heads are hidden by the ashtray or the glove compartment door. After you've removed all visible screws, gently try to pry the panel off. Sometimes a putty knife is helpful here. If you find a spot that doesn't budge, you've either missed a screw or there's a spring clip behind it. Don't use too much force or you'll break the trim panel. Crutchfield MasterSheets can really come in handy here because they show you the locations of hidden clips and screws unique to your vehicle.

After you've taken off the trim panel, remove all the screws attaching the factory head unit to the dash and gently try to pull it out. If it seems stuck, there may be some type of rear support. Once you slide the old head unit out, disconnect the antenna plug and factory wiring harness.

If you're installing a GM- or Chrysler-style aftermarket head unit, the installation process is exactly the reverse of the removal process.

If you're installing a Euro DIN, ISO-DIN, or shaft-style head unit, you'll need a mounting adapter kit. Most adapter kits are combination Euro DIN and shaft style or combination ISO-DIN and shaft style, where pieces are cut away or omitted for nonshafted applications.

For a Euro DIN head unit, the first step is to screw the adapter to the dash. It should mount the same way the original head unit did. Next, slide the metal mounting sleeve supplied with the head unit (Fig. 5-12) into the adapter opening until the small outer lip makes contact with the edges of the opening. Once you put the sleeve in position, bend the tabs on the sleeve out to secure it in place. Connect the wires to the head unit and test it. Now push the head unit into the sleeve—it should click into place. Install the trim ring, factory trim panel, and backstrap.

For an ISO-DIN head unit, you'll first need to mount the head unit to the adapter. This is achieved by using screws in matching holes in the adapter and sides of the head unit. Next mount the adapter to the dash. Install the factory trim panel and backstrap.

Figure 5-12
Tabs on Euro DIN
sleeve. (*Courtesy of
Crutchfield.*)

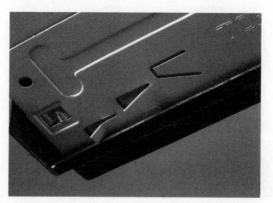

For a shaft-style head unit, the best approach is to install the head unit into the mounting adapter first. Put the faceplate included with the head unit on in this step. Next mount the adapter to the dash. Install the factory trim panel and backstrap.

Removing and Installing in a Euro DIN Dash Opening. A Euro DIN dash opening gives you the option of installing either a Euro DIN or a shaft-style head unit. Euro DIN installations are usually quick and easy. Removal and installation is done from the front, and no dash panels need to be removed.

The factory head unit is secured in a sleeve by spring clips, which you release using a pair of DIN removal tools (Fig. 5-13). These look like U-shaped pieces of bent coat hanger and are often included with a new head unit. Most receivers also have some type of rear support that needs to be detached before the old head unit is pulled out. Once you slide the old head unit out, disconnect the antenna plug and factory wiring harness.

Your new Euro DIN head unit will come with a metal mounting sleeve. You may be able to use the sleeve already in place in the dash, or you may need to replace it with the new one. See if your new head unit slides in properly and whether it locks into place securely in the old sleeve. If not, remove the original sleeve by bending back the metal tabs holding it in place and sliding it out. Slide the new sleeve into position until the outer lip makes contact with the edges of the opening. Make sure the top of the sleeve is facing up, then bend the tabs on the sleeve out to secure it in place.

After you make the appropriate wiring connections and plug in the antenna, slide the new head unit into the sleeve. Try the head unit to make sure it's functioning properly. If everything is OK, push it all the way in until it clicks into place.

Figure 5-13
DIN removal tools in action. (*Courtesy of Crutchfield.*)

Installing a shaft-style head unit in a Euro DIN opening requires a simple conversion kit (Fig. 5-14). The kit provides a shaft-style mounting plate and hardware to attach it to a Euro DIN dash opening.

Removing and Installing in an ISO-DIN Dash Opening. ISO-DIN dash openings are common among Toyotas and Nissans. The ISO-DIN mounting method uses factory side support brackets, so no mounting adapter is needed for ISO-DIN head units. A shaft-style head unit can also be mounted in an ISO-DIN dash using a suitable mounting adapter.

Before you can install your new head unit, you'll need to remove the old one. Normally a factory trim panel will need to be removed first. This might be attached with screws or spring clips or a combination of the two. Often, screw heads are hidden by the ashtray or the glove compartment door. After you've removed all visible screws, gently try to pry the panel off. Sometimes a putty knife is helpful here. If you find a spot that doesn't budge, you've either missed a screw or there's a spring clip behind it. Don't use too much force or you'll break the trim panel.

After you've taken off the trim panel, remove all the screws attaching the factory head unit assembly to the dash and gently try to pull it out. If it seems stuck, there may be some type of rear support. Once you slide the old head unit assembly out, disconnect the antenna plug and factory wiring harness.

For an ISO-DIN head unit, remove the ISO-DIN brackets from the sides of the factory head unit and install them to the sides of the new head unit in the same way (Fig. 5-15). The plastic trim ring included with the new head unit for Euro DIN installations will not be used here. From this point, the installation process is exactly the reverse of the removal process. Install the head unit assembly in the dash, then install the factory trim panel and backstrap.

Figure 5-14
Scosche DIN-to-shaft conversion kit DIN1190. (*Courtesy of Scosche.*)

Figure 5-15
Mounted ISO-DIN brackets. (*Courtesy of Crutchfield.*)

Tip: Many cassette and CD head units are shipped with a plastic or metal locking screw that you'll need to remove before installation. The owner's manual will tell you whether this is the case.

For a shaft-style head unit, you'll need to use a mounting adapter. The best approach is to install the head unit into the mounting adapter first. Put the faceplate included with the head unit on in this step. Next mount the adapter to the dash. Install the factory trim panel and backstrap.

Antenna Mounting. If you're installing an antenna in a vehicle that didn't have one before, scout out the antenna location and cable path the car's designers intended. Look for a rubber or plastic plug on the front fender. Another likely location on Japanese cars is the front pillar between the windshield and front door. Do not run the antenna cable through a hole with other wiring or you will risk picking up interference from those wires.

In the unlikely event that you don't find an existing mounting hole, you'll need to drill one. Check where other cars of the same make have theirs. Once you've chosen a location, protect the area with a piece of duct tape. Use a center punch to mark it and drill a ⅛-inch pilot hole. Gradually enlarge the hole with successively larger bits to reduce the chance of slipping. Check the hole as you approach the correct size to avoid drilling too large a hole.

Electrical Wiring

The key to making the electrical wiring part of head unit installation easy is to use the proper wiring harness and, if necessary, original equipment manufacturer (OEM) integration and antenna adapters.

Wiring Harness. Snapped into the back of your factory head unit, you will find one or more rectangular wiring plugs. (If your car was ordered with a radio prep package but no radio, these plugs will be dangling behind the dash where the radio would go.) This is called the *factory wiring harness,* and it contains power, speaker, and various other wires such as a power antenna turn-on.

Unfortunately, the factory wiring harness will not directly connect to an aftermarket head unit. For most vehicles, however, you can purchase a wiring harness (Fig. 5-16) to mate with the existing car harness, so you don't need to splice into any of the factory wires. (Crutchfield includes one of these harnesses with the purchase of a head unit.) The wires of this mating harness are first connected to the appropriate wires of the aftermarket head unit. The other end of the harness then snaps right into your car's factory wiring harness.

There are a number of good reasons to use the mating wiring harness approach (Fig. 5-17). With a mating harness, you can do the job of connecting wires on a well-lit workbench rather than underneath a dashboard. Leaving the factory wiring harness intact makes it easy if you ever decide to reinstall the factory radio. And when you buy a mating wiring harness, the job of identifying each of the factory wires is done for you.

Use butt-splice or closed-end-splice crimp connectors to attach the wiring harness wires to your new head unit's wires. If you prefer, you can

Figure 5-16
World of wiring harnesses. (*Courtesy of Scosche.*)

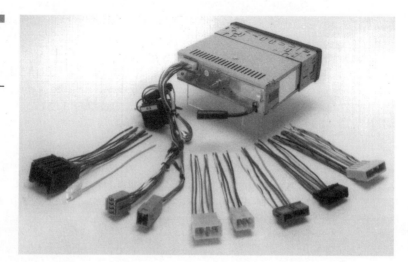

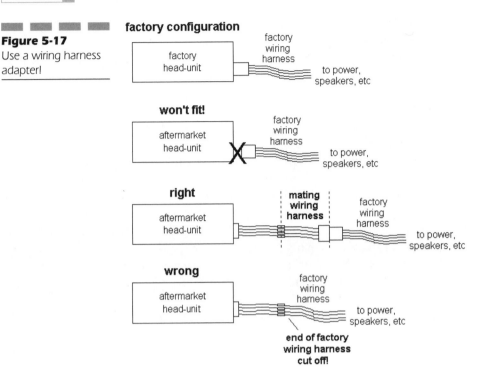

Figure 5-17
Use a wiring harness
adapter!

solder the wire ends together and use heat-shrink tubing. Some harnesses come with bullet connectors already attached, to mate with those on the head unit's wires. The most important factor is getting a tight connection that won't come loose over time. Harnesses are supplied with a color code reference chart identifying all the wires (or each of the wires is labeled), so you don't need to figure them out yourself.

If a wiring harness isn't readily available, Scotchlok connectors can be used to connect an aftermarket head unit to a factory wiring harness without cutting off the end of the harness. This makes it easy if you ever decide to reinstall the factory radio.

OEM Integration Adapters. In most cases, a wiring harness is all you need to properly connect an aftermarket head unit to a vehicle's power and speaker wires. However, there are many cases where you need to use an OEM integration adapter between the head unit and wiring harness. These include:

- Premium factory sound systems
- Vehicles with chassis-grounded speakers
- Vehicles with common-grounded speakers

How do you know if you're a candidate for an OEM integration adapter? The easiest and most reliable method is to consult with the pros—stop by your favorite local car stereo shop or call Crutchfield. They should each have access to an up-to-date database containing the information. You may be able to scope it out yourself. Check for evidence of a premium factory system by looking for logos on the head unit or speaker grilles. Chassis-grounded and common-grounded speakers can often be detected by inspection of trunk speaker wiring.

If you have a premium factory sound system (such as Delco/Bose or Ford/JBL), the section on premium factory sound systems later in this chapter has details about suitable OEM integration adapters.

Vehicles with chassis-grounded speakers or common-grounded speakers not connected to chassis ground present a problem for high-power head units. This is because the minus speaker wires from high-power head units cannot be grounded or tied together and are needed to carry music as well as DC voltage to each of the speakers.

With chassis-grounded or common-grounded speakers, you actually have three options:

- Use a low-power head unit.
- Rewire the system so two independent wires are used for each speaker.
- Use an OEM integration adapter.

A low-power head unit is an option only if you can live with the limited volume.

Rewiring the system is a lot of work, but has the added benefit of letting you put in speaker wire better than the 22-gauge stuff commonly used for factory speaker wiring. If you're willing to spend the time, this is the best option from a performance standpoint.

OEM integration adapters provide a quick and easy solution to the problem of using high-power head units in many vehicles. Unfortunately, there may be performance sacrifices involved. In the case of OEM integration adapters for chassis-grounded speaker wiring, it is normal to simply AC-couple the plus speaker wires through a capacitor and not use the minus speaker wires from the head unit at all. This results in slashing the

■■ ■■ ■■ ■■

Figure 5-18
SoundGate FLT1.
(*Courtesy of
SoundGate.*)

power to the speakers by a factor of 4—no better than using a low-power head unit!

In the case of OEM integration adapters for common-grounded (but not connected to chassis ground) speaker wiring, it is possible to maintain a high-power output. Although not all units do this, one device that does is the FLT1 by SoundGate (Fig. 5-18). Not all OEM integration adapters of this type maintain a full 20- to 20,000-Hz frequency response, so check before you buy.

Antenna Adapters.   Aftermarket head units all use the standard Motorola antenna connector, but many factory antennas do not. That's where antenna adapters come in. Adapters are available to adapt various factory antennas to aftermarket head units (Table 5-7). Adapters are also available to adapt aftermarket antennas to factory head units, so be careful to get the right adapter.

TABLE 5-7

Types of Factory
Antenna
Connectors

Type	Comments
Standard (Motorola)	Used on all aftermarket equipment.
GM mini	Used on 1988 and up GM.
Ford	Used on some 1995 and up vehicles, usually for radios with DSP controls.
Nissan	Diversity (two-antenna) system.
Volkswagen/Euro	Factory antenna needs 12 volts to turn on impedance-matching circuit.

Factory Steering Wheel Control Interfaces

Many late-model vehicles have steering wheel audio controls in addition to the normal controls on the head unit. They make it easier to control your stereo without taking your hands off the wheel or your eyes off the road. When you install an aftermarket head unit, you normally lose use of these auxiliary controls.

SoundGate now offers solutions to this problem with their factory steering wheel control interfaces (Fig. 5-19). These allow you to retain use of auxilliary head unit controls with Sony head units having RMX-2S or RMX-4S wired remote capabilities.

The GMSW1 works with the factory steering wheel controls of most late-model GM vehicles. The FRDSW1 works with the factory "driver convenience" dashboard controls as well as the rear-seat controls of most Ford vehicles.

Premium Factory Sound Systems

If you have a premium factory sound system (such as Delco/Bose or Ford/JBL) and want to replace your head unit with an aftermarket model, you'll need to use a suitable OEM integration adapter between the head unit and wiring harness. This section explains the basics of how pre-

Figure 5-19
Soundgate FRDSW1.
(*Courtesy of
SoundGate.*)

mium factory sound systems work and how to choose the right OEM integration adapter.

Premium Factory Sound System Basics

Although the details of Bose and Ford Premium systems differ (as do different Bose systems among themselves) they share the same basic principle. They all use external amplifiers (amplified speakers or an outboard amp) and they all use differential signals between the head unit and amplifiers.

The advantage of using differential signals is noise immunity. The differential outputs of the head unit provide two equal but opposite versions of the music signal (an inverted version and a noninverted version). At the differential inputs of the amplifier, the two versions of the signal are subtracted. Since any externally induced noise presumably affects both versions of the signal the same, subtraction causes cancellation of the noise. Because one of the music signals was inverted, subtraction also results in a music signal that is twice the original signal.

The differential outputs of the head unit have a DC offset to eliminate the need for coupling capacitors in the head unit and amplifiers. DC offset is also used in Bose systems to "wake up" the amplifiers.

Bose systems often incorporate speakers with impedances much less than the usual 4 ohms (0.4 ohms is common!). This allows the Bose amplifiers to achieve much higher power levels without having to incorporate expensive DC-to-DC converters to provide a higher supply voltage.

Chrysler/Infinity Systems

Like Bose and Ford Premium systems, Chrysler/Infinity systems use external amplifiers (amplified speakers). Unlike Bose and Ford, Chrysler/Infinity systems use standard head unit technology. This approach allows upgrading to the premium system in the assembly plant by simply using amplified speaker assemblies instead of normal ones.

Some earlier Chrysler/Infinity systems (prior to 1995) use low-power head units; the rest use high-power head units. If you want to replace a high-power factory head unit with a high-power aftermarket model (and keep the amplified speakers), no OEM integration adapter is required. If the factory head unit is low power, you will need to use an OEM integration adapter of the type normally used for chassis-grounded speaker wiring.

If you're replacing the factory head unit, be sure to connect the remote turn-on wire from the aftermarket head unit to the remote turn-on wire in the factory harness. Chrysler/Infinity systems use this wire to "wake up" the amplifiers.

Another approach you can use when replacing the factory head unit is to bypass the amplifiers at the amplified speakers. This is feasible in Chrysler/ Infinity systems because standard-impedance speakers are used. (This is not the case for Bose systems, which commonly use ultra-low-impedance speakers.) By bypassing, you will lose the benefit of custom equalization built into the factory amps, but you will gain the ability to boost the head unit with your own amp.

Choosing the Right OEM Integration Adapter

The biggest decision you'll need to make when shopping for an OEM integration adapter is whether to buy a universal model or a vehicle-specific one (see Table 5-8).

Tip: Many universal interfaces are two-channel units. If this is the case, you will need to buy two units to retain functionality of your head unit fader and balance controls.

Universal, "one size fits all" premium factory sound adapters were the norm in the eighties, when only a few premium factory sound system variations existed. This type of interface offers the advantages of generally lower cost and better availability—retailers only need to stock one model. This type of adapter is usually called an OEM interface differential converter and requires gain adjustments for optimum performance.

Vehicle-specific premium factory sound adapters are designed specifically for the model and year of your vehicle. These are becoming the trend as the number of system variations explodes. Vehicle-specific interfaces come with the proper wiring harness already attached, so you don't need to figure out which wires go where. They eliminate the need for any

TABLE 5-8

Universal Versus Vehicle-Specific Adapters

	Type	Comments
	Universal	Generally lower cost and better availability—retailers only need to stock one model.
		Requires gain adjustments for optimum performance.
	Vehicle-specific	Designed specifically for the model and year of your vehicle.
		Comes with the proper wiring harness already attached, so you don't need to figure out which wires go where.
		Eliminates the need for any type of adjustment.

Figure 5-20
SoundGate Premium
Sound System
Interface. (*Courtesy of*
SoundGate.)

Figure 5-20
SoundGate Premium
Sound System
Interface. (*Courtesy of*
SoundGate.)

type of adjustment because they have been optimized for each vehicle. Vehicle-specific interfaces are customarily four-channel units. SoundGate manufactures adapters of this type (Fig. 5-20).

A vehicle-specific interface is no guarantee of quality in itself, but when a manufacturer has gone to the trouble to research each specific vehicle, there is an added assurance that everything will work as advertised.

Amplifiers and Amplifier Projects

With today's high-power (13 watts per channel and up) head units, you may not need separate amps to provide the volume levels you want. But if you have a low-power head unit you want to keep or want to play extra loud, then separate amps are what you need.

In addition to boosting head units, separate amps are used for adding subwoofers and biamping. Because of their complexity, the latter projects are given their own chapters, but be sure to read this chapter for important information on choosing and using amps.

The project sections in this chapter focus on using amps to boost all types of head units. Head units are categorized according to their available outputs, and fall into one of the first five project classes shown in Table 6-1.

Before jumping into the head unit boosting projects, there are a few things you should know about amplifiers.

Advertised Power versus Honest Power

Unlike the home stereo industry, the car stereo industry has not adopted a uniform standard of power measurement. By fudging the power supply voltage and distortion limit of the measurement, car stereo manufacturers are able to claim inflated power ratings. Table 6-2 shows how this

TABLE 6-1

Amplifier Projects

Project	Comments
Boosting head units with four preamp outputs and four speaker outputs	
Boosting head units with two preamp outputs and four speaker outputs	
Boosting head units with no preamp outputs and four speaker outputs	
Boosting head units with no preamp outputs and two speaker outputs	
Boosting premium factory sound system head units	
Adding a subwoofer	See Chap. 4
Biamping	See Chap. 8

TABLE 6-2

Amplifier Power
Rating Methods

Measurement Method	Load	Supply Voltage	THD	Power Rating
Honest power	4 ohms	13.8 volts	0.1%	13 watts per channel
RMS power	4 ohms	14.4 volts	1%	18 watts per channel
Maximum power	4 ohms	14.4 volts	10%	25 watts per channel
Peak power	4 ohms	14.4 volts		35 watts per channel

works, rating an amplifier using various methods. The amplifier is an 18-watt per channel amplifier made by a reputable manufacturer.

Honest power (my terminology) uses 13.8 volts for the supply voltage, a typical value for most cars when the engine is running. It also uses an audibly acceptable distortion limit of 0.1 percent. This limit is similar to that used by the home stereo industry. For this amplifier, the power rating using this method is 13 watts per channel.

RMS power uses a 14.4-volt supply and allows 1 percent distortion. This allows manufacturers to claim higher power levels than the honest power method. For this amplifier, the power rating using RMS power is 18 watts per channel. The RMS power method is the closest thing to a standard for the reputable segment of the industry.

Maximum power uses a 14.4-volt supply and allows 10 percent distortion. This allows manufacturers to claim even higher power levels than with the RMS power method. For this amplifier, the power rating using the maximum power method is 25 watts per channel.

An even more inflated method of specifying power is the peak power method. This method measures the instantaneous power at the peak output voltage of the amplifier. For this amplifier, the power rating using this method would be 35 watts per channel—almost 3 times the honest power value!

The point of all this is that you should make sure you are comparing apples to apples when you are shopping for amps or head units. Look at the measurement conditions specified. If none are specified, assume the worst. I have seen flea market amps use 15-volt supply voltages and 3.2-ohm loads in addition to the peak power measurement method to achieve spectacular power ratings for low-power amps.

How Much Power Do You Need?

Two good rules of thumb regarding amplifier power are:

- Doubling the volume level requires 10 times more amplifier power.
- A just noticeable increase in volume level requires a 10 percent increase in amp power.

This can be good news or bad news depending on how you look at it. If you have 50 watts per channel now and want to be able to play twice as loud, you'll need 500 watts per channel. On the other hand, you can have half the volume level of a 500-watt per channel amp by buying a 50-watt per channel model.

The bottom line is, achieving high volume levels becomes increasingly expensive in terms of amplifier power.

How much power you need depends on how loud you want to play your music. Calibrate yourself by listening to a high-powered head unit in your vehicle or a similar one. Drive at highway speeds with the window cracked for the worst-case situation. High-powered head units are normally rated at 13 to 18 watts per channel, so adjust this figure according to the rules of thumb. If you need a little less than twice the volume, then 100 watts per channel would be about right. Speaker sensitivity also comes into play here; you'll need more power for inefficient speakers and less for efficient ones. Remember, 3 dB lower speaker sensitivity translates to needing twice the amplifier power to provide the same volume level.

More Power for Your Money

Table 6-3 shows two tricks you can sometimes use to get more watts per dollar from separate amplifiers.

TABLE 6-3

Methods of Getting More Watts per Dollar from Amps

Technique	Comments
Run 2-ohm loads	Must use two 4-ohm speakers wired in parallel or a dual voice coil subwoofer
Run bridged amps	Bridging with four-channel amps usually provides the best value

Running 2-Ohm Loads

The formula for power (from your amp to your speakers) is:

$$\text{Power} = \frac{\text{Amplifier Output Voltage}^2}{\text{Speaker Impedance}}$$

Virtually all car speakers are 4 ohms. By using 2-ohm speakers, you can double the power from the amp! It sounds too good to be true. It is. There are three limitations to this scheme:

- You must use an amp that is 2-ohm stable.

- You don't actually double the power—a factor of 1.5 is more typical. This is because the amplifier output voltage drops when driving a 2-ohm load.

- Two-ohm speakers are almost impossible to come by—you must use two 4-ohm speakers wired in parallel or a dual voice coil subwoofer.

The unavailability of 2-ohm speakers is the big problem with this idea for most installations. Using two speakers in parallel does make sense if you need multiple speakers per channel to handle unusually high power levels. Parallel front/rear speaker configurations also take advantage of this trick. This wiring is shown in Fig. 6-1.

A more common situation for running 2-ohm loads occurs with dual voice coil subwoofers. Subwoofers are commonly available with dual 4-ohm voice coils. There is little or no cost premium for the additional voice coil. These can be wired in parallel to create a 2-ohm subwoofer as shown in Fig. 6-2.

Tip: An amp driving 2-ohm loads runs hotter, so be sure to mount it where it will have good air circulation.

Bridging

Another method that can sometimes give you more watts per dollar is called *bridging.*

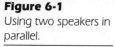

Figure 6-1
Using two speakers in parallel.

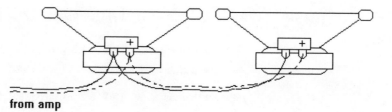

from amp

Figure 6-2
Parallel-wired DVC
subwoofer.

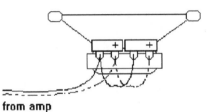

from amp

Bridging converts a stereo amp into a mono amp, which gives you more power than the sum of the two stereo channels. Bridging typically provides 3 times the power of a single amplifier channel (which is 1.5 times the combined power of two stereo channels). As an example, a 50-watt per channel stereo amp might be bridgeable to a 150-watt mono configuration. Similarly, a 50-watt per channel four-channel amp might be bridgeable to a 150-watt per channel stereo configuration.

There are four limitations to this scheme:

■ You must use an amp that is bridgeable.

■ You need twice as many amplifier channels.

■ A bridged amp cannot normally drive less than 4 ohms.

■ The lowest bridged power is about 75 watts per channel.

Whether or not bridging can give you more watts per dollar depends on several factors. Suppose you need two channels of 150 watts for a set of 4-ohm speakers. Table 6-4 shows what you might find.

By bridging model A, you get 150 watts, but you would need to buy two amps for a grand total of $280. Model B is the conventional solution—a two-channel, 150-watt per channel amp for $230. This is usually cheaper than buying two model As. The third option is to bridge model C—for this example, this ends up being the cheapest solution. Any of these options could have been the most cost effective, depending on what's on sale. But in general, four-channel amps not only start out as a cheaper approach, they are more aggressively marked down when clearance time comes around!

Tip: An amp running in the bridged mode runs hotter, so be sure to mount it where it will have good air circulation.

TABLE 6-4

Bridging Cost
Comparison

Model	Power	Bridged Power	Price Each	Total
A	50W × 2 channels	150W × 1 channel	$140	$280
B	150W × 2 channels		$230	$230
C	50W × 4 channels	150W × 2 channels	$200	$200

You must normally use a 4-ohm or higher speaker load when bridging. This means you can't run two 4-ohm woofers wired in parallel, for example. This is because a bridged amplifier "sees" a speaker load of half the actual value, so a 4-ohm speaker looks like a 2-ohm equivalent load. Very few amps can handle using anything less than 4 ohms in the bridged mode. So if you were thinking of saving big money by running 2-ohm loads *and* bridging with the same amplifier, forget it.

Be sure to follow the instructions included with the amp for the correct bridging procedure and connections—not all amps are alike.

Where to Put Your Amps

Where to put your amps is more a matter of convenience than performance, but you should always mount amps away from noise sources and in a spot with good air circulation. This means avoiding locations like underneath the driver's side of the dash and near the blower motor. The most popular mounting locations are shown in Table 6-5.

TABLE 6-5

Popular Amplifier Mounting Locations

	Mounting Location	Comments
BEST FOR FRONT SPEAKER AMPS	Under dash (passenger side only)	Short wires and patch cords can be used
		Space is tight
		Moderate electrical noise
BEST FOR REAR SPEAKER AMPS	Trunk or hatch area	Plenty of space, near rear speakers
		Low electrical noise location
		Long patch cords from head unit may be needed
		Subject to damage from sliding cargo
	Under seat	Short wires and patch cords can be used
		Space is very tight
		Subject to dirt and spills

Important: When mounting a power amp (or any car stereo component), secure it in such a way that the metal case does not make electrical contact with the car chassis. This guarantees that the amplifier's ground wire is the exclusive connection between the amp's internal ground and the system ground.

NOTE   A few components have electrically isolated cases—these should be electrically connected to the car chassis to provide shielding to the internal circuitry. When in doubt, use an ohmmeter to measure the resistance between the case and ground wire of the component.

Use rubber grommets inside metal mounting bracket holes to prevent screws from making an electrical connection with the chassis. Alternately, you can mount your components on a board, then mount the board to the car chassis.

Electrical isolation is important to prevent ground loops, which are a major source of noise problems in car audio. See Chap. 11 for more on this subject.

If you plan to use a four-channel amp to drive both front and back speakers, then the best location is probably under the passenger side of the dash—if it will fit. Many four-channel amps are too big for this location.

Power and Ground Wiring

Power and ground wiring is one of those subjects a lot of people don't pay much attention to. But improper wiring can easily rob your amplifiers of 25 percent or more of their rated power—or, worse yet, cause a fire.

Wire Gauge

There are two factors that determine the recommended wire gauge for amplifier power and ground wires:

- Voltage drop
- Wire overheating

Voltage drop is determined by the resistance of the power wire multiplied by the current. It results in the amplifier getting less than the full battery voltage for power. A drop of just 1 volt can result in a 15 percent

Voltage, Current, Resistance, and Power

The best way to understand the four basic electrical quantities is by thinking about water and plumbing:

voltage (volts) = water pressure (pounds per square inch)

current (amps) = water flow (gallons per minute)

resistance (ohms) = pipe resistance

power (watts) = pressure × flow

Voltage and current are easy to understand—voltage is like water pressure and current is like water flow.

Electrical resistance is analogous to a pipe's resistance to flow. Imagine trying to fill a swimming pool with a regular garden hose. Then think about how much longer it would take if the hose were the diameter of spaghetti. This is because of the higher resistance of the spaghetti-sized hose. In both cases the water pressure is the same, but in the case of the higher resistance, the flow is reduced. With electricity, this same relationship holds true. It's called Ohm's law:

current = voltage/resistance

This says that you can double the current (water flow) by doubling the voltage (water pressure) or by halving the resistance (pipe resistance). Ohm's law can be rearranged to calculate any of the three quantities if you know the other two.

In electricity, power is calculated using this equation:

power = voltage × current

In terms of water, this means that you need both water pressure and water flow to hose the mud off your car. Having lots of pressure but little flow is like trying to wash the car with a squirt gun. Having lots of flow but little pressure is like gently pouring buckets of water over the car.

This equation can be rearranged and combined with Ohm's law to let you calculate any of the four quantities if you know two of them.

loss of maximum power to the speakers. Because of the high currents consumed by power amps, voltage drops can be significant.

Take a 100W × 2 amp, for example. It can provide 200 watts of power to a pair of speakers, but draws about 30 amps to do so, because of efficiency losses in the amp. Suppose you need to run 20 feet of power wire to the amp in the trunk. Using 10-gauge wire (1.0 milliohms per foot), the wire resistance is 20 milliohms. With 30 amps of current, you would get a voltage drop of 0.60 volts, resulting in roughly a 10 percent loss of maximum power to the speakers. Using 8-gauge wire (0.64 milliohms per foot) reduces this to a more acceptable 6 percent loss of maximum power.

Wire overheating is a concern because of the high currents drawn by amplifiers. The current-carrying capacity of a wire is determined by its resistance per foot and the acceptable temperature rise in the wire. Exceeding the current rating of a wire will cause melting of the insulation and result in possible short circuits or fire.

Table 6-6 shows the recommended wire gauge, depending on wire length, for various power amps. The wire lengths shown in the table guarantee a drop of less than 0.5 volts in the power wire. This should result in less than 8 percent loss of maximum amplifier power. Using a thicker wire (lower gauge) and shorter length will always give you better performance. The wire gauges shown in Table 6-6 also guarantee a temperature rise of less than 35°C in the wire at maximum current. NEVER use thinner wire (higher gauge) than that shown in Table 6-6, even for short wires, or you risk a meltdown. For example, #10 is the thinnest wire you should use with a $100W \times 2$ amp, even for a 1-foot length of wire.

If you are running multiple amps on the same power wire, just add their power ratings together. For example, a pair of $50W \times 2$ amps should use the $100W \times 2$ row of the table.

TABLE 6-6

Recommended Wire Gauge for Power Amps

Amplifier	Power Wire		Amplifier	Power Wire	
$20W \times 2$	Up to 13'	#18	$100W \times 2$	Up to 17'	#10
	Up to 20'	#16		Up to 26'	#8
	Up to 32'	#14	$150W \times 2$	Up to 17'	#8
$35W \times 2$	Up to 12'	#16		Up to 28'	#6
	Up to 18'	#14	$200W \times 2$	Up to 21'	#6
	Up to 30'	#12		Up to 33'	#4
$50W \times 2$	Up to 13'	#14	$350W \times 2$	Up to 19'	#4
	Up to 21'	#12		Up to 30'	#2
	Up to 33'	#10	$500W \times 2$	Up to 21'	#2
$75W \times 2$	Up to 14'	#12		Up to 26'	#1/0
	Up to 22'	#10			
	Up to 35'	#8			

Power Wire Hookup

If you are only adding a single 18W × 2 amp, then you can get away with using the factory distribution/fuse block for power. Otherwise, because of the high currents drawn by power amps, you will need to provide your own power supply wiring directly to the battery. Figures 6-3 and 6-4 show two methods of doing this.

Figure 6-3 is a diagram of a single wire running through the firewall. A fuse block is located very close to the battery to protect against shorts on the battery side of the distribution/fuse block. The distribution/fuse block may be located under the dash, in the trunk, or in any other convenient location.

> **Important:** The wire from the battery to the distribution/fuse block must have a current rating high enough for the total current drawn by all the amps.
>
> To determine the proper wire gauge for this wire, add together the power ratings of all the amps, then refer to the previous table. For example, a 100W × 2 amp plus a 50W × 2 amp should use the 150W × 2 row of the table.
>
> For a pair of identical amps, you may simply choose a wire gauge 4 less than the highest gauge (smallest wire) recommended for a single amp. For example, if you have a pair of 50W × 2 amps, use #10 wire to the battery since #14 wire is the highest gauge recommended for a 50W × 2 amp.

Figure 6-4 is a diagram of multiple wires running through the firewall. The distribution/fuse block is located very close to the battery to protect against shorts. This approach eliminates the need for a separate fuse block, but you have to run multiple wires through the firewall.

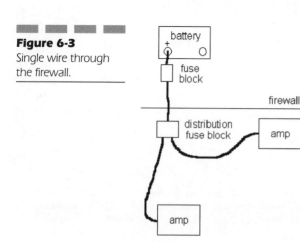

Figure 6-3
Single wire through the firewall.

Figure 6-4
Multiple wires through
the firewall.

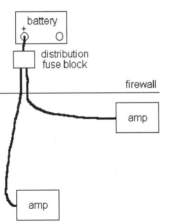

You may be able to use an existing hole in the firewall to run wires from the engine compartment to the passenger compartment. Otherwise, you'll need to drill a hole in a strategic location. Always protect the wires from the sharp metal edges of the hole by using a rubber grommet or split loom tubing.

Fuses

Even choosing the right fuse requires some thought. In addition to the standard AGC glass tube and ATO/ATC blade styles, there are giant versions of each of these for high-current applications (called AGU and MAXI, respectively). For ultra-high-current applications, there are two types of wafer-style fuses. Figure 6-5 shows a variety of fuse styles.

The main factor in deciding which style of fuse to use is the current rating. For applications up to 35 or 40 amps, use AGC or ATO/ATC. For up to 60 or 80 amps, use AGU or MAXI. Above 80 amps, you must use a wafer-type fuse. Table 6-7 provides further comparison of the different fuse types.

Choose a fuse holder that accommodates the wire gauge you plan to use. High-amperage fuse blocks are generally designed for 8-gauge or 4-gauge wire. Distribution fuse blocks usually have a 4-gauge input and multiple 8-gauge outputs.

> **Important:** Make sure the fuse holder is rated to handle the current of the fuse you plan to use, or you can melt or burn the holder. Do not assume a fuse holder is designed to work safely with any fuse that will fit in it. For example, many AGC fuse holders are only rated at 10 amps, even though AGC fuses can go up to 35 amps.

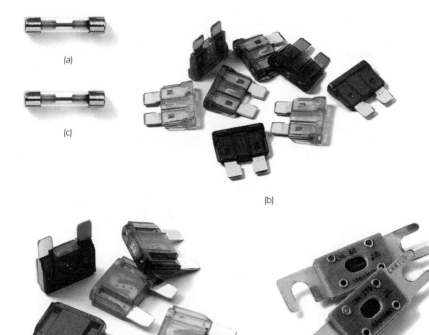

Figure 6-5
(a) AGC fuse; (b) ATO/ATC fuses; (c) AGU fuse; (d) MAXI fuses; (e) ANL fuses; (f) MEGA fuse.
(*Courtesy of Littelfuse.*)

For high-current situations, using circuit breakers instead of fuses is attractive because of the high cost of replacement fuses. In many cases, the cost of a circuit breaker is actually less than the cost of a fuse plus fuse holder. Since circuit breakers do not require holders, you can actually save money even if you never overload a circuit. A disadvantage with circuit breakers is that if you need to change the current rating, it is more difficult than just popping in a different fuse.

TABLE 6-7

Fuse Types for Auto Sound

Fuse Type	Description	Current	Pricing	Comments
AGC	Glass tube $1\frac{1}{4} \times \frac{1}{4}$	up to 35 amps	Fuse $.50 Holder $2	Standard automotive fuse of the seventies
ATO/ATC	Blade	up to 40 amps	Fuse $.50 Holder $2	Standard automotive fuse of the eighties
AGU	Glass tube $1\frac{1}{2} \times \frac{13}{32}$	up to 60 amps	Fuse $1 Holder $7—16	Large version of AGC type
MAXI	Blade	20—80 amps	Fuse $1 Holder $4—11	Large version of ATO/ATC
ANL	Wafer (bolt-on)	60—250 amps	Fuse $6—14 Holder $8—40	Required for very high-current applications
MEGA	Wafer (bolt-on)	100—250 amps	Fuse $13—15 Holder $25	Required for very high-current applications

Calculating Fuse Current Rating

The simplest method to determine the correct current rating for a fuse is to add together the maximum currents of everything on that circuit. The owner's manuals should list the maximum current; otherwise, look at the equipment fuses.

An alternate rule of thumb method is to add the rated power of all amplifier channels together and divide by 6. For example, for a pair of 30W × 2 amps (four channels altogether):

$$(30 + 30 + 30 + 30)/6 = 20$$

In this case you would use a 20-amp fuse.

Wiring for Automatic Turn-On

Most amps have a remote turn-on lead—the amp turns on when 12 volts is applied to this lead. Normally, you connect this lead to the power antenna lead of your head unit (or amp turn-on lead, if your head unit provides one). Such a connection will turn your amp on and off with your head unit.

If your head unit doesn't have a wire suitable for amplifier turn-on (some Ford/Lincoln/Mercury vehicles provide a 5-volt signal, some power antenna leads only work when you actually use the radio, other head units may have no wire at all), you'll have to use the ignition/accessory 12 volts instead.

Another option is to buy a remote power adapter. This device monitors the DC voltage on a convenient speaker wire (5 to 7 volts for a high-power head unit) and provides a 12-volt amp turn-on signal to the amp. Even if your head unit has a suitable turn-on wire, this device can save you from having to run a wire from the head unit to the trunk.

If you have multiple amplifiers in your system, you can usually tie the turn-on leads together. If you have four or more amps in your system, the power turn-on lead of the head unit may not be able to provide enough current, and you risk damaging the head unit.

A relay can solve this problem. It's connected to your system as shown in Fig. 6-6.

The head unit turn-on lead energizes the magnetic coil inside the relay, which activates the internal switch. The switch connects the amp remote turn-on leads to 12 volts. Here are the key points for success:

■ Use a 12-volt automotive relay or relay with 12VDC coil. The coil rating should be 100 mA or less to prevent damaging the head unit. The contact rating should be 2 amps or higher.

■ Connect the relay coil terminals to the head unit turn-on lead and ground as shown. The two coil terminals may or may not be interchangeable—the relay case will show + and − if they are not.

■ Connect the relay switch terminals as shown. Use ignition/accessory 12 volts or battery 12 volts. Some relays have both NO (normally open) and NC (normally closed) switch terminals in addition to COM (common). Use the NO and COM terminals. They may be interchanged.

Ground Wire Hookup

The metal chassis of a car forms a huge, low-impedance ground plane. The basic grounding strategy is to ground each amplifier (or other component) at the nearest chassis ground location.

Figure 6-6
Using a relay for more current drive.

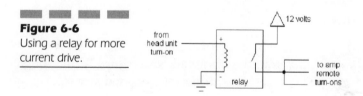

The ground wire should be the same gauge as the power wire and must make direct contact with bare metal of the car. Not all metal in a car is electrically connected to the chassis, so use a multimeter to confirm a low-resistance connection to a known good ground.

Look for an existing bolt or screw that makes contact with the car body near the amp. Remove the bolt or screw and scrape away any paint or dirt. A star washer will help your ground wire maintain good contact with the car body. Crimp and solder a ring terminal onto the end of the ground wire, slip it underneath the washer, and replace the bolt or screw. If you can't find a convenient ground screw or bolt, drill a hole for one. Be careful not to drill into the gas tank or a gas or brake line.

If you're installing several components, ground the amps separately. This prevents the high currents drawn by an amp from modulating the ground voltage of any of the other components. It's okay for low-current components to share a common ground point.

> **Important:** Some cars (Audi, Porsche) have galvanized bodies. In these cars, you must find one of the manufacturers' grounding points or else some noise can result.

Connectors

Using the proper wire gauge is an important factor in getting the power you paid for from your amps. Another equally important factor is connectors.

After about six months, a properly crimped connection has an impedance of roughly 0.03 ohms, due to oxidation (it gets even worse as time goes on). This doesn't sound like much, but consider the fact that from the battery to the amp there can easily be five or more crimped connections through distribution blocks and so forth.

Let's assume five connections of 0.03 ohms, for a grand total of 0.15 ohms. For a 50W × 2 amp, the maximum current is approximately 17 amps. This results in a maximum voltage drop of 2.5 volts, and a reduction in maximum power of over 30 percent. Remember, this is due to the connectors alone, and is on top of any additional losses in the wires themselves.

Crimping and then soldering results in an impedance of 0.01 ohms per connection. This number does not degrade over time. For our example, this would limit the maximum voltage drop to 0.85 volts, and the reduction in maximum power to 10 percent.

The point of this is:

■ Solder crimped connectors where high currents are involved. This means power supply and ground wires for all power amps. While you're at it, soldering crimped speaker wire connectors is a good idea too.

■ Eliminate unnecessary connections.

Soldering crimped spade lugs and quick disconnects is simple because the end of the wire is exposed. For high-current butt splices, don't use crimp connectors. Solder the ends directly and insulate with heat-shrink tubing.

Eliminating unnecessary connections favors using the "multiple wires through the firewall" power wire hookup approach explained previously.

Speaker Wiring

For speaker wiring, the dominating consideration in selecting wire gauge is damping factor. Damping factor is the ratio of speaker impedance to driving impedance, where driving impedance is usually dominated by the speaker wires. For 4-ohm speakers, the speaker wires should have an impedance of less than 0.1 ohms for best performance. Impedance values higher than this will result in damping factors less than 40, which can start to affect bass response.

Table 6-8 provides minimum recommended wire gauges, depending on wire length, for all power amps. The wire lengths shown in Table 6-8 guarantee damping factors greater than 40. Using thicker wire (smaller gauge) and shorter lengths will give you higher damping factors.

If you are running 2-ohm loads, subtract 2 from the recommended wire gauge in the chart. For example, use #10-2 up to 30 feet.

Boosting Head Units with Four Preamp Outputs and Four Speaker Outputs

Head units with four preamp outputs provide the most flexibility for boosting with amplifiers. The preamp outputs provide a low-noise, low-

TABLE 6-8

Minimum Recommended Gauge for Speaker Wire

Amplifier	Speaker Wire	
All wattages	up to 12′	#16-2
	up to 20′	#14-2
	up to 30′	#12-2

distortion interface to any amplifier. Having two sets of preamp outputs means the head unit fader control can remain functional for all but one of the configurations.

There are five configurations with this type of head unit. The first two configurations let you provide as much power as you want to each of two pairs of speakers. The third configuration lets you save an amp when equal power to front and rear speakers is acceptable. The last two configurations let you save an amp when high power is needed for only one pair of speakers. This may seem like a strange idea at first, but it actually can make a lot of sense. Table 6-9 shows the five configurations.

Standard Configuration

The standard configuration boosts all four channels. Two stereo amps or a single four-channel amp may be used. In this configuration, the speaker outputs of the head unit are unused.

You may choose to use two stereo amps with different power ratings if more power is needed for one pair of speakers. One example of this is for rear fill applications, where the rear speakers are operated at a reduced volume level to improve imaging (see the "Imaging Overhaul" section in Chap. 3). Another example is where the front speakers have much lower power handling capacity than the rear speakers. In this case, you might want to run a more powerful amp in back and put bass blocking filters at the preamp-level inputs of the front amp. Use 100- or 200-Hz, 12-dB/octave filters for best results. (Many amps now include built-in 100-Hz high-pass filters.) These prevent deep bass frequencies from wasting amplifier power and distorting your front speakers, so you can run a high volume level up front. Since deep bass is nondirectional, the back speakers will fill in the missing deep bass in front.

Cable-Saving Configuration

This configuration lets you provide as much power as you want to two pairs of speakers.

If you choose to use a trunk-mounted amp for the rear speakers, this configuration saves having to run a patch cord to the trunk-mounted amp since the speaker wires already in place are used to carry the signal to the amp. Either the trunk amp will need to be a model that accepts speaker-level inputs or you will need to use a line output converter to convert the speaker-level signals to preamp level (see box on p. 179).

TABLE 6-9

Configurations for
Boosting Head
Units with Four
Preamp Outputs
and Four Speaker
Outputs

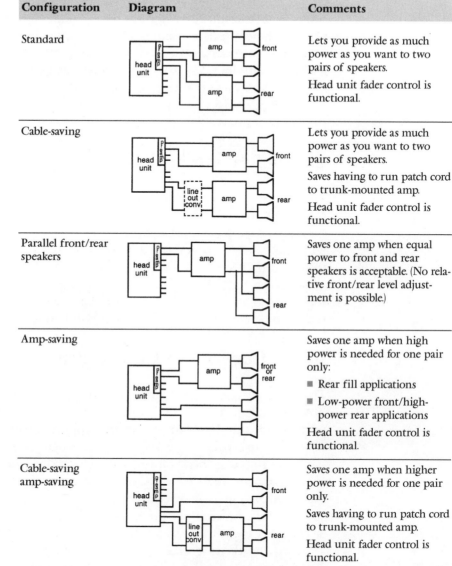

Configuration	Diagram	Comments
Standard		Lets you provide as much power as you want to two pairs of speakers. Head unit fader control is functional.
Cable-saving		Lets you provide as much power as you want to two pairs of speakers. Saves having to run patch cord to trunk-mounted amp. Head unit fader control is functional.
Parallel front/rear speakers		Saves one amp when equal power to front and rear speakers is acceptable. (No relative front/rear level adjustment is possible.)
Amp-saving		Saves one amp when high power is needed for one pair only: ▪ Rear fill applications ▪ Low-power front/high-power rear applications Head unit fader control is functional.
Cable-saving amp-saving		Saves one amp when higher power is needed for one pair only. Saves having to run patch cord to trunk-mounted amp. Head unit fader control is functional.

You may choose to use two stereo amps with different power ratings if more power is needed for one pair of speakers. One example of this is for rear fill applications, where the rear speakers are operated at a reduced volume level to improve imaging (see the "Imaging Overhaul" section in Chap. 3). Another example is where the front speakers have much lower power handling capacity than the rear speakers. In this case, you might want to run a more powerful amp in back and put bass blocking filters at the preamp-level inputs of the front amp. Use 100- or 200-Hz, 12-dB/octave filters for best results. (Many amps now include built-in 100-Hz high-pass filters.) These prevent deep bass frequencies from wasting amplifier power and distorting your front speakers, so you can run a high volume level up front. Since deep bass is nondirectional, the back speakers will fill in the missing deep bass in front.

For proper fader operation, be sure to use the correct set of head unit speaker outputs.

Parallel Front/Rear Speakers Configuration

The parallel front/rear speakers configuration shares a single amp for front and rear channels. This gives you almost the same power per speaker as having identical separate amps for front and rear. The drawback with this configuration is that you are stuck with having equal power for front and back speakers.

Since the amplifier used in this configuration will see a parallel speaker load, it must be 2-ohm stable. An amp driving 2-ohm loads runs hotter, so be sure to mount it where it will have good air circulation.

Amplifier-Saving Configuration

The amplifier-saving configuration is the same as the standard configuration, except that one amplifier is eliminated by using an existing amp in the head unit. This can be useful in applications where more power is needed for one pair of speakers than the other.

One example of this is for rear fill applications, where the rear speakers are operated at a reduced volume level to improve imaging (see the "Imaging Overhaul" section in Chap. 3). Another example is where the front speakers have much lower power handling capacity than the rear speakers. In this case, you could use the head unit speaker outputs in front.

Be sure to use the correct set of preamp outputs for this configuration. The fader control will work properly for one set and oppositely for the other set.

Cable-Saving Amplifier-Saving Configuration

This configuration is useful when higher power is needed for only the rear pair of speakers and you want to mount the amp in the trunk. Either the amp will need to be a model that accepts speaker-level inputs, or you will need to use a line output converter to convert the speaker-level signals to preamp level.

As with the previous configuration, this one lets you save an amp and maintain fader functionality. By using the existing rear speaker wires to carry the signal to the amp, you save having to run a patch cord from the head unit to the trunk-mounted amp.

Boosting Head Units with Two Preamp Outputs and Four Speaker Outputs

There are five configurations with this type of head unit. The first two let you provide as much power as you want to each of two pairs of speakers. The third lets you save an amp when equal power to front and rear speakers is acceptable. The last two configurations let you save an amp when high power is needed for only one pair of speakers. The cable-saving version of the amp-saving configuration saves having to run a patch cord to a trunk-mounted amp when you are boosting the rear speakers. Table 6-10 shows the five configurations.

Standard Configuration

In this configuration, Y cables are used to feed the preamp-level outputs from the head unit to two stereo amps (or a single four-channel amp). The speaker outputs of the head unit are unused. You may choose to use two stereo amps with different power ratings if more power is needed for one pair of speakers. One example of this is for rear fill applications, where the rear speakers are operated at a reduced volume level to improve imaging (see the "Imaging Overhaul" section in Chap. 3). Another example is where the front speakers have much lower power handling capacity than the rear speakers. In this case, you might want to run a more powerful amp in back and put bass blocking filters at the preamp-level inputs of the front amp. Use 100- or 200-Hz, 12-dB/octave filters for best results. (Many amps now include built-in 100-Hz high-pass filters.) These prevent deep bass frequencies from wasting amplifier power and distorting your front speakers, but still let you run a high volume level up front. Since deep bass is nondirectional, the back speakers will fill in the missing deep bass in front.

Set the relative front/rear level by using the gain controls on the amps.

TABLE 6-10

Configurations for
Boosting Head
Units with Two
Preamp Outputs
and Four Speaker
Outputs

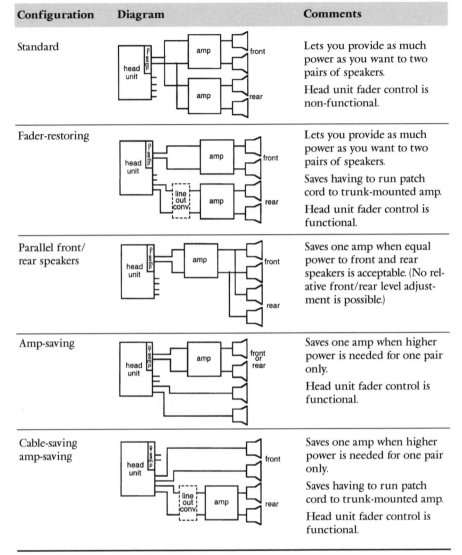

Configuration	Diagram	Comments
Standard		Lets you provide as much power as you want to two pairs of speakers. Head unit fader control is non-functional.
Fader-restoring		Lets you provide as much power as you want to two pairs of speakers. Saves having to run patch cord to trunk-mounted amp. Head unit fader control is functional.
Parallel front/ rear speakers		Saves one amp when equal power to front and rear speakers is acceptable. (No relative front/rear level adjustment is possible.)
Amp-saving		Saves one amp when higher power is needed for one pair only. Head unit fader control is functional.
Cable-saving amp-saving		Saves one amp when higher power is needed for one pair only. Saves having to run patch cord to trunk-mounted amp. Head unit fader control is functional.

Fader-Restoring Configuration

This configuration lets you provide as much power as you want to two pairs of speakers, and restores use of the fader. This is accomplished by using speaker-level outputs from the head unit as the inputs to the second amp. Either the second amp will need to be a model that accepts speaker-level inputs, or you will need to use a line output converter to convert the speaker-level signals to preamp level.

If you choose to use a trunk-mounted amp for the rear speakers, this configuration saves having to run a patch cord to the trunk-mounted amp since the speaker wires already in place are used to carry the signal to the amp.

You may choose to use two stereo amps with different power ratings if more power is needed for one pair of speakers. One example of this is for rear fill applications, where the rear speakers are operated at a reduced volume level to improve imaging (see the "Imaging Overhaul" section in Chap. 3). Another example is where the front speakers have much lower power handling capacity than the rear speakers. In this case, you might want to run a more powerful amp in back and put bass blocking filters at the preamp-level inputs of the front amp. Use 100- or 200-Hz, 12-dB/octave filters for best results. (Many amps now include built-in 100-Hz high-pass filters.) These prevent deep bass frequencies from wasting amplifier power and distorting your front speakers, but still let you run a high volume level up front. Since deep bass is nondirectional, the back speakers will fill in the missing deep bass in front.

For proper fader operation, be sure to use the correct set of head unit speaker outputs.

Parallel Front/Rear Speakers Configuration

The parallel front/rear speakers configuration shares a single amp for front and rear channels. This gives you almost the same power per speaker as having identical separate amps for front and rear. The drawback with this configuration is that you are stuck with having equal power for front and back speakers.

Since the amplifier used in this configuration will see a parallel speaker load, it must be 2-ohm stable. An amp driving 2-ohm loads runs hotter, so be sure to mount it where it will have good air circulation.

Amplifier-Saving Configuration

This configuration lets you save an amp when higher power is needed for only one pair of speakers.

One example of this is for rear fill applications, where the rear speakers are operated at a reduced volume level to improve imaging (see the "Imaging Overhaul" section in Chap. 3). Another example is where the front speakers have much lower power handling capacity than the rear speakers. In this case, you could use the head unit speaker outputs in front.

For proper fader operation, be sure to use the correct set of head unit speaker outputs.

Cable-Saving Amplifier-Saving Configuration

This configuration is useful when higher power is needed for only the rear pair of speakers and you want to mount the amp in the trunk. Either the amp will need to be a model that accepts speaker-level inputs, or you will need to use a line output converter to convert the speaker-level signals to preamp level.

As with the previous configuration, this one lets you save an amp and maintain fader functionality. By using the existing rear speaker wires to carry the signal to the amp, you save having to run a patch cord from the head unit to the trunk-mounted amp.

Boosting Head Units with No Preamp Outputs and Four Speaker Outputs

Head units without preamp outputs can be boosted by choosing amplifiers that accept speaker-level inputs. Alternately, line output converters can be used to convert a head unit's speaker-level signals to preamp level for use with any amp.

There are four configurations for head units with no preamp outputs and four speaker outputs. The standard configuration provides the most flexibility, and lets you provide as much power as you want to each of two pairs of speakers. The converter-saving configuration can eliminate the need for a second line output converter (at the expense of losing fader control). The parallel front/rear speaker configuration lets you save an amp when equal power to front and rear speakers is acceptable. The amp-saving configuration lets you save an amp when high power is needed for only one pair of speakers. This may seem like a strange idea at first, but it actually can make a lot of sense. Table 6-12 shows the four configurations.

Standard Configuration

The standard configuration boosts all four channels and retains the functionality of the head unit fader control. If you use amps without speaker-level inputs, two line output converters are required for this configuration. Two stereo amps or a single four-channel amp may be used.

Line Output Converters

Line output converters reduce the signal amplitude from speaker level to preamp level. Better models also eliminate common-mode noise and break ground loops.

Table 6-11 shows the three basic types of line output converters.

TABLE 6-11

Types of Line Output Converters

Type	Price	Active or Passive	Comments
Resistive	$2–$30	Passive	*Does not* eliminate common-mode noise or break ground loops. Typically very small.
Transformer	$15–$30	Passive	Eliminates common-mode noise and breaks ground loops. Typically palm-sized case.
⚡BEST⚡ Differential amp	$15–$30	Active	Eliminates common-mode noise and breaks ground loops. Typically palm-sized case.

Resistive converters are the least expensive, but do not provide ground isolation. Transformer and differential amplifier converters do provide ground isolation, but cost more. Differential amp converters provide better frequency response and are less susceptible to magnetic fields than transformer converters.

Ground isolation helps reduce system noise problems caused by ground loops. Some amplifiers provide ground isolation at their preamp-level inputs, in which case there is no additional benefit to having isolation in the line output converter. The level of benefit provided by ground isolation also depends on the ground structure of your installation. Trunk-mounted amps usually benefit the most from ground isolation.

When deciding what to buy, either buy a differential amp model right off the bat or start with a $2 resistive converter and upgrade if necessary.

You may choose to use two stereo amps with different power ratings if more power is needed for one pair of speakers. One example of this is for rear fill applications, where the rear speakers are operated at a reduced volume level to improve imaging (see the "Imaging Overhaul" section in Chap. 3). Another example is where the front speakers have much lower power handling capacity than the rear speakers. In this case, you might want to run a more powerful amp in back and put bass blocking filters at the preamp-level inputs of the front amp. Use 100- or 200-Hz, 12-dB/octave

TABLE 6-12

Configurations for
Boosting Head
Units with No
Preamp Outputs
and Four Speaker
Outputs

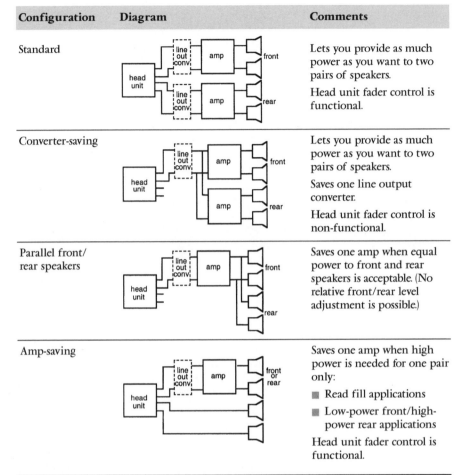

Configuration	Diagram	Comments
Standard		Lets you provide as much power as you want to two pairs of speakers. Head unit fader control is functional.
Converter-saving		Lets you provide as much power as you want to two pairs of speakers. Saves one line output converter. Head unit fader control is non-functional.
Parallel front/ rear speakers		Saves one amp when equal power to front and rear speakers is acceptable. (No relative front/rear level adjustment is possible.)
Amp-saving		Saves one amp when high power is needed for one pair only: ■ Read fill applications ■ Low-power front/high-power rear applications Head unit fader control is functional.

filters for best results. (Many amps now include built-in 100-Hz high-pass filters.) These prevent deep bass frequencies from wasting amplifier power and distorting your front speakers, but still let you run a high volume level up front. Since deep bass is nondirectional, the back speakers will fill in the missing deep bass in front.

Converter-Saving Configuration

The converter-saving configuration is similar to the standard configuration, except that a line output converter is eliminated by sharing a single converter for both amps. If you use amps with speaker-level inputs, then there's no advantage to this configuration. The disadvantage is that the fader control on the head unit becomes nonfunctional. You can still set the relative front/rear level by using the controls on the amps themselves, but you lose the convenience of head unit control.

Parallel Front/Rear Speakers Configuration

The parallel front/rear speakers configuration shares a single amp for front and rear channels. This gives you almost the same power per speaker as having identical separate amps for front and rear. The drawback with this configuration is that you are stuck with having equal power for front and back speakers.

Since the amplifier used in this configuration will see a parallel speaker load, it must be 2-ohm stable. An amp driving 2-ohm loads runs hotter, so be sure to mount it where it will have good air circulation.

Amplifier-Saving Configuration

The amplifier-saving configuration is similar to the standard configuration, except that one amplifier is eliminated by using an existing amp in the head unit. This can be useful in applications where more power is needed for one pair of speakers than the other.

One example of this is for rear fill applications, where the rear speakers are operated at a reduced volume level to improve imaging (see the "Imaging Overhaul" section in Chap. 3). Another example is where the front speakers have much lower power handling capacity than the rear speakers. In this case, you could use the external amp in back.

Boosting Head Units with No Preamp Outputs and Two Speaker Outputs

There are four configurations for head units with no preamp outputs and two speaker outputs (Table 6-13). The first simply boosts two speakers. The last three allow you to add a set of speakers.

TABLE 6-13

Configurations for
Boosting Head
Units with No
Preamp Outputs
and Two Speaker
Outputs

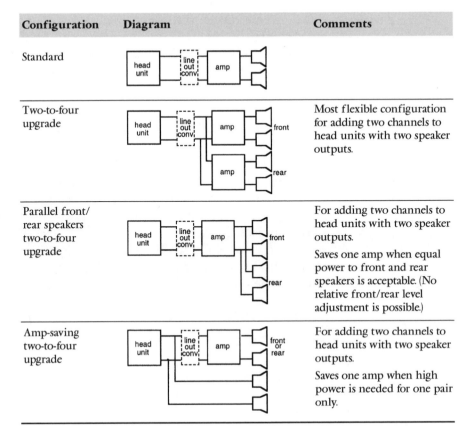

Configuration	Diagram	Comments
Standard		
Two-to-four upgrade		Most flexible configuration for adding two channels to head units with two speaker outputs.
Parallel front/ rear speakers two-to-four upgrade		For adding two channels to head units with two speaker outputs. Saves one amp when equal power to front and rear speakers is acceptable. (No relative front/rear level adjustment is possible.)
Amp-saving two-to-four upgrade		For adding two channels to head units with two speaker outputs. Saves one amp when high power is needed for one pair only.

Standard Configuration

The standard configuration boosts two channels. If you use an amp without speaker-level inputs, a line output converter is required for this configuration.

Two- to Four-Speaker Upgrade Configuration

This configuration lets you boost the original speakers as well as add a second pair of boosted speakers.

You may choose to use two stereo amps with different power ratings if more power is needed for one pair of speakers. One example of this is for rear fill applications, where the rear speakers are operated at a reduced volume level to improve imaging (see the "Imaging Overhaul" section in Chap. 3). Another example is where the front speakers have much lower power handling capacity than the rear speakers. In this case, you might want to run a more powerful amp in back and put bass blocking filters at the preamp-level inputs of the front amp. Use 100- or 200-Hz, 12-dB/octave filters for best results. (Many amps now include built-in 100-Hz high-pass filters.) These prevent deep bass frequencies from wasting amplifier power and distorting your front speakers, but still let you run a high volume level up front. Since deep bass is nondirectional, the back speakers will fill in the missing deep bass in front.

Set the relative front/rear level by using the controls on the amps.

Parallel Front/Rear Speakers Configuration

The parallel front/rear speakers configuration lets you add a pair of speakers by sharing a single amp for front and rear channels. This gives you almost the same power per speaker as having identical separate amps for front and rear. The drawback with this configuration is that you are stuck with having equal power for front and back speakers.

Since the amplifier used in this configuration will see a parallel speaker load, it must be 2-ohm stable. An amp driving 2-ohm loads runs hotter, so be sure to mount it where it will have good air circulation.

Amplifier-Saving Two- to Four-Speaker Upgrade Configuration

This configuration lets you add a pair of speakers and takes advantage of the existing amp in the head unit to drive the other, lower-power pair of speakers. This is useful for rear fill applications, where the rear speakers are operated at a reduced volume level to improve imaging (see the "Imaging Overhaul" section in Chap. 3). Another example is where the front speakers have much lower power handling capacity than the rear speakers. In this case, you could use the external amp in back.

Because the head unit lacks a fader control, the proper front/rear level balance must be set by adjusting the level control on the amp.

Boosting Premium Factory Sound System Head Units

If you have a premium factory sound system (such as Delco/Bose or Ford/JBL) and want to boost your head unit with an amp, the situation is more complicated than with standard systems. You'll need to bypass the factory amps plus use a line output converter designed specifically for Bose or Ford Premium systems. Depending on the system, you may need to replace the factory speakers as well.

Premium Factory Sound System Basics

Although the details of Bose and Ford Premium systems differ (as do different Bose systems among themselves), they share the same basic principle. All use external amplifiers (amplified speakers or an outboard amp) and all use differential signals between the head unit and amplifiers.

The advantage of using differential signals is noise immunity. The differential outputs of the head unit provide two equal but opposite versions of the music signal (an inverted version and a noninverted version). At the differential inputs of the amplifier, the two versions of the signal are subtracted. Since any noise presumably affects both versions of the signal the same, subtraction causes cancellation of the noise. Because one of the music signals was inverted, subtraction also results in a music signal that is twice the original signal.

The differential outputs of the head unit have a DC offset to eliminate the need for coupling capacitors in the head unit and amplifiers. DC offset is also used in Bose systems to "wake up" the amplifiers.

Bose systems often incorporate speakers with impedances much less than the usual 4 ohms (0.4 ohms is common!). This allows the Bose amplifiers to achieve much higher power levels without having to incorporate expensive DC-to-DC converters to provide a higher supply voltage.

Chrysler/Infinity Systems

Like Bose and Ford Premium systems, Chrysler/Infinity systems use external amplifiers (in the form of amplified speakers). Unlike Bose and Ford, Chrysler/Infinity systems use standard head unit technology. This approach allows upgrading of the premium system in the assembly plant by simply using amplified speaker assemblies instead of normal ones.

If you want to boost a Chrysler/Infinity head unit with an aftermarket amp, you will need to bypass the factory amps. By bypassing, however, you will lose the benefit of custom equalization built into the factory amps. Chrysler/Infinity systems use standard-impedance speakers, so you will not need to replace the speakers if you decide to drive them with an aftermarket amp.

Some earlier Chrysler/Infinity systems (prior to 1995) used low-power head units; the rest use high-power head units. In either case, a standard line output converter will do the job of interfacing a Chrysler/Infinity head unit to an amp lacking speaker-level inputs.

Do You Need to Replace Your Factory Speakers?

Most Bose systems use speakers that are only 0.4 ohms (instead of the usual 4 ohms). Aftermarket amps can't handle driving this low of an impedance. If your vehicle uses low-impedance speakers, you'll need to replace them if you want to boost your head unit with an aftermarket amp.

It's smart to scope this out beforehand, so you know what you're getting into. Remove one of the factory speaker assemblies. Disconnect the wires going to the driver itself, then use a multimeter to measure the resistance of the driver. If it's less than 3.2 ohms, you'll need to replace the speakers with aftermarket ones.

An easier alternative is to call Crutchfield. They can tell you what you've got and how to proceed. They also stock a wide selection of adapters and harnesses for interfacing to premium factory sound systems.

Bypassing Factory Amps

Bose and Ford Premium sound systems use external amplifiers (in the form of amplified speaker assemblies or outboard amps). These amplifiers need to be bypassed in order to use an aftermarket amp.

In the case of amplified speaker assemblies, the aftermarket amp should be located near the head unit to make best use of the factory wiring. Each speaker assembly will need to be removed and modified to bypass its amp. For amplified speaker assemblies with low-impedance factory speakers (less than 4 ohms), replace the entire amplified speaker with a 4-ohm aftermarket speaker. For amplified speaker assemblies with 4-ohm factory speakers, bypass the signal wires feeding each amplifier to the speaker itself.

For outboard factory amp systems, the aftermarket amp can be located near the head unit or at the factory amp location, to make best use of the factory wiring. In either case, the wiring harness from the output of the factory amp will need to be unplugged from the factory amp and driven by the aftermarket amp.

Line Output Converters for Bose/Ford Premium Head Units

Line output converters are commonly used to convert head unit speaker-level outputs to levels suitable for the preamp-level inputs of amps.

If you're dealing with a Bose or Ford Premium Sound factory head unit, you'll want to use a line output converter specifically designed for interfacing to those units. Using a standard LOC having low-input imped-ance may damage the output circuit of a Bose or Ford Premium Sound head unit. Using a standard LOC with a Bose or Ford system can also result in extremely low output from the LOC because of the relatively low signal levels of many premium factory system head units.

The LOCB by SoundGate (Fig. 6-7). will accept any input signal from 175 mV to 6.6 volts and convert it to a 2.5-volt audio output suitable for aftermarket crossovers and amps. It also has a noise-blanking circuit that activates the audio path after turn-on noises have subsided.

Tip: Most line output converters are two-channel units. If this is the case, you will need to buy two con-verters to retain func-tionality of your head unit fader and bal-ance controls.

Figure 6-7
SoundGate LOCB.
(*Courtesy of SoundGate.*)

System Configurations

There are four configurations for boosting premium factory sound system head units. The standard configuration provides the most flexibility, and lets you provide as much power as you want to each of two pairs of speakers. The converter-saving configuration can eliminate the need for a second line output converter (at the expense of losing fader control). The parallel front/rear speaker configuration lets you save an amp when equal power to front and rear speakers is acceptable. The amp-saving configuration lets you save an amp when high power is needed for only one pair of speakers. This may seem like a strange idea at first, but it actually can make a lot of sense. Table 6-14 shows the four configurations.

TABLE 6-14

Configurations for Boosting Premium Factory Sound System Head Units

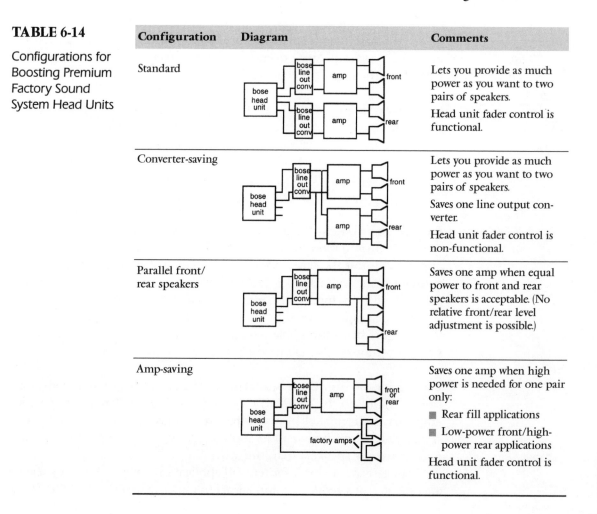

Configuration	Diagram	Comments
Standard		Lets you provide as much power as you want to two pairs of speakers. Head unit fader control is functional.
Converter-saving		Lets you provide as much power as you want to two pairs of speakers. Saves one line output converter. Head unit fader control is non-functional.
Parallel front/rear speakers		Saves one amp when equal power to front and rear speakers is acceptable. (No relative front/rear level adjustment is possible.)
Amp-saving		Saves one amp when high power is needed for one pair only: ■ Rear fill applications ■ Low-power front/high-power rear applications Head unit fader control is functional.

Standard Configuration.   The standard configuration boosts all four channels and retains the functionality of the head unit fader control. Two line output converters are required for this configuration. Two stereo amps or a single four-channel amp may be used.

You may choose to use two stereo amps with different power ratings if more power is needed for one pair of speakers. One example of this is for rear fill applications, where the rear speakers are operated at a reduced volume level to improve imaging (see the "Imaging Overhaul" section in Chap. 3). Another example is where the front speakers have much lower power handling capacity than the rear speakers. In this case, you might want to run a more powerful amp in back and put bass blocking filters at the preamp-level inputs of the front amp. Use 100- or 200-Hz, 12-dB/octave filters for best results. (Many amps now include built-in 100-Hz high-pass filters.) These prevent deep bass frequencies from wasting amplifier power and distorting your front speakers, but still let you run a high volume level up front. Since deep bass is nondirectional, the back speakers will fill in the missing deep bass in front.

Converter-Saving Configuration.   The converter-saving configuration is similar to the standard configuration, except that a line output converter is eliminated by sharing a single converter for both amps. The disadvantage is that the fader control on the head unit becomes nonfunctional. You can still set the relative front/rear level by using the controls on the amps themselves, but you lose the convenience of head unit control.

Parallel Front/Rear Speakers Configuration.   The parallel front/rear speakers configuration shares a single amp for front and rear channels. This gives you almost the same power per speaker as having identical separate amps for front and rear. The drawback with this configuration is that you are stuck with having equal power for front and back speakers.

Since the amplifier used in this configuration will see a parallel speaker load, it must be 2-ohm stable. An amp driving 2-ohm loads runs hotter, so be sure to mount it where it will have good air circulation.

Amplifier-Saving Configuration.   The amplifier-saving configuration is similar to the standard configuration, except that an aftermarket amplifier is eliminated by retaining use of factory amps for one pair of speakers. This can be useful in applications where more power is needed for one pair of speakers than the other.

One example of this is for rear fill applications, where the rear speakers are operated at a reduced volume level to improve imaging (see the "Imaging Overhaul" section in Chap. 3). Another example is where the front speakers have much lower power handling capacity than the rear speakers. In this case, you could use the aftermarket amp in back.

Adjusting Amp Gain Settings

Once you've installed your amp or amps, you'll need to set the level controls for optimum performance. If you set the level too low, you'll have noise problems. If you set the level too high, you'll have distortion problems. The goal is to find the level that maximizes the signal-to-noise ratio while avoiding of distortion.

Follow these steps, depending on the configuration of your system.

Procedure for Two-Amp Configurations

1. Set the amp level controls to minimum and the head unit fader control to its center detent position. If you're using any line output converters with their own level controls, set them to minimum.

2. With music, turn up the head unit volume control (you should barely be able to hear the music). When you reach the point where you start to hear distortion from any speaker, turn the volume control down slightly until the distortion disappears. Leave the volume control at this setting.

NOTE *If the sound from the speakers is so low that you cannot clearly hear the onset of distortion, turn up the amp's level controls a bit.*

3. If you're using any line output converters with their own level controls, return all amp level control settings to minimum, then turn up each LOC level control until you reach a point where you start to hear distortion. Back the LOC level controls down slightly until the distortion disappears. (If you never reach a point where you hear distortion, the LOC level controls should be set to max.)

4. Turn up your front amp level control until the system is as loud as you'll ever play it or you begin to hear distortion. If you hear distortion, slightly decrease the amp level control.

5. Turn down the head unit volume to a normal listening level.

6. Turn up the rear amp level until the desired front/rear balance is achieved.

Procedure for Amp-Saving Configurations

If you're not using a line output converter with its own level controls, skip to Step 4.

1. Set the amp level controls and LOC level controls to minimum. Set the head unit fader control all the way toward the channels with the amp.

2. With music, turn up the head unit volume control (you should barely be able to hear the music). When you reach the point where you start to hear distortion from a speaker (on an amplified channel), turn the volume control down slightly until the distortion disappears. Leave the volume control at this setting.

NOTE   *If the sound from the speakers is so low that you cannot clearly hear the onset of distortion, turn up the amp's level controls a bit.*

3. Return all amp level control settings to minimum, then turn up each LOC level control until you reach a point where you start to hear distortion. Back the LOC level controls down slightly until the distortion disappears. (If you never reach a point where you hear distortion, the LOC level controls should be set to max.) Turn down the head unit volume control before proceeding to step 4.

4. Set the head unit fader control to its center detent position.

5. With music, adjust the head unit volume to a normal listening level.

6. Adjust the amp level controls until the desired front/rear balance is achieved.

Equalizers and Equalizer Projects

To Equalize or Not?

Although equalizers were once thought of as the savior of great sound quality for high-end systems, the trend for their use is down. The philosophy of many seems to be that less is more. In other words, the fewer the number of components, the better the guard against the introduction of noise and distortion. Better-quality speakers, good speaker placement, and subwoofers have alleviated much of the need for equalizers.

On the other hand, the acoustics of cars are notorious for introducing frequency response peaks and valleys. Even if your speakers are perfect, your overall frequency response will be bumpy. Boomy-sounding bass due to resonances, as well as irregularities in the mid-range and treble due to window reflections, are typical. Equalization still offers the best solution to these problems.

How Many Bands Do You Need?

It's important to differentiate between the different classes of equalizers (see Fig. 7-1).

Five- and seven-band graphic equalizers are essentially sophisticated tone controls, and don't give you the control you need to correct the narrow peaks and valleys of your car's acoustic response. On the positive side, these definitely give you more flexibility than just bass and treble controls, and can be set reasonably well by ear. For this reason they are best mounted within easy reach.

Thirteen- through thirty-band graphic equalizers, and most parametric equalizers, can be considered tools. They give you the control you need to correct real-world response problems. Unfortunately, they are more expensive, and cannot reliably be set by ear—a ⅓-octave real-time analyzer (RTA) is required to set them properly. The good news is that any

Figure 7-1
Classes of equalizers.

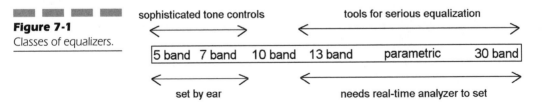

sophisticated tone controls   tools for serious equalization

| 5 band | 7 band | 10 band | 13 band | parametric | 30 band |

set by ear   needs real-time analyzer to set

serious car stereo shop should have one of these, and should be willing to help you set your equalizer for a reasonable fee. Equalizers falling into the tools category are normally set once, then left alone. For this reason, they are best mounted out of sight and out of reach of curious fingers. Be sure to write down the settings of each control or use a marker on the face to indicate the proper setting.

Ten-band graphic equalizers fall in the transition region between sophisticated tone controls and tools. Table 7-1 shows a summary of the performance levels for various equalizers.

The EQL Series II equalizer by AudioControl is a thirteen-band model with a twist. In the critical bass region, $\frac{1}{2}$-octave equalization is provided. Above that, octave equalization are used. For most purposes, this is like getting a twenty-band ($\frac{1}{2}$ octave) equalizer for the price of a thirteen-band model. This equalizer is clearly in the tool category. It sells for $229.

The EQT Series II equalizer by AudioControl (Fig. 7-2) is a thirty-band ($\frac{1}{3}$-octave) model. It's a favorite among competitors, but outside the budget of most noncompetitors. A pair of these (you'll need two, since they're single channel) will set you back $600.

Parametric equalizers have been around a long time, but have never enjoyed the popularity of graphic equalizers. Instead of the familiar slider controls at each frequency band, parametric equalizers have center, bandwidth, and gain controls. This provides plenty of flexibility to combat response peaks and valleys, but makes setting these equalizers quite difficult without a real-time analyzer.

TABLE 7-1

Performance Levels for Various Equalizer Types

Type	Comments
Five-band	Better than tone controls, but inadequate for serious equalization.
Seven-band	Better than tone controls, but inadequate for serious equalization.
Ten-band (octave)	Can correct broad response problems.
Thirteen-band (octave with $\frac{1}{2}$-octave bass)	Can correct most response problems—see text.
Thirty-band ($\frac{1}{3}$-octave)	The ultimate!
Parametric	Can correct most response problems, depending on number of stages.

Figure 7-2
EQT Series II 30-band
equalizer. (*Courtesy of
AudioControl.*)

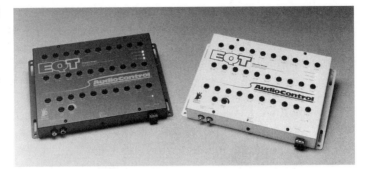

What About Equalizer Boosters?

Equalizer boosters typically combine a seven-band graphic equalizer with an 18W × 2 channel or 18W × 4 channel power amp. This was a popular package in the eighties, but seems to be dying out. The decrease in popularity of equalizers in general combined with the high power of most modern head units is one reason. Another is the deservedly poor reputation of these units for high noise and distortion.

The prices of equalizer boosters are amazingly low, but they're a good example of "you get what you pay for."

Connecting Your Equalizer

Figure 7-3 shows the standard connection for a preamp-level (often incorrectly called passive) equalizer.

Some models provide speaker-level as well as preamp-level inputs. Since equalizers are normally two-channel devices, the head unit fader control is rendered inoperative. Dash-mounted equalizers often provide their own fader controls and separate front and rear outputs to remedy this problem. When using an equalizer without a fader, the relative front/rear level will need to be set by adjusting the amp level controls.

How to Set Your Equalizer

You can set equalizers of ten bands or less fairly well by ear, but you need a real-time analyzer to set thirteen- to thirty-band or parametric equaliz-

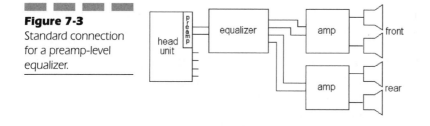

Figure 7-3
Standard connection
for a preamp-level
equalizer.

ers properly. Note: If you do have access to an RTA, it can improve the quality of your five- to ten-band equalizer settings as well.

Setting by Ear

Initial Preparation. Start by setting your head unit bass and treble controls flat, and your equalizer sliders to their center detent positions. The fader control should be in the position you normally use when listening. Pick a CD or tape that you like and that sounds good on your home stereo. Choose one that covers the entire audio spectrum—something with bass drums, voice, and cymbals. Acoustic instruments such as piano, sax, and so on are better than synthesized ones because you know what they should sound like. Play your selection at a normal listening level.

Familiarize yourself with how each of the frequency bands sounds. One by one, slide a control to the max position and listen to the effect, then return it to the center position. Try to characterize each frequency band in your head, for example, "solid bass," "boomy bass," "female vocals," and so on. This will help you later as you try to compensate for things like bass that is too boomy or weak female vocals.

Setting the Bass. The area that normally benefits the most from equalization is the bass. Car acoustics tend to overemphasize mid-bass, so start by reducing the level in the 200-Hz region. For a seven-band equalizer, this means moving the 150-Hz slider down a third of the way. For a ten-band equalizer, move the 120- and 250-Hz sliders down a fourth of the way. You will probably want to boost the bass in the 60-Hz region to provide a more solid bottom end. This will also help overcome the masking effects of road noise. Don't overdo it—maybe a third of the way up is good.

Leave the 30-Hz slider of a ten-band equalizer set to its center detent position. Boosting 30 Hz will not give you much benefit (few speaker systems can reproduce this range effectively): It will instead dramatically reduce your amp headroom and cause your woofers to distort on those passages that have energy in this frequency range.

Adjusting by Ear.   Now the fun part. One by one, slide each control up and down a little bit and listen to the effect. Ask yourself if the music sounds more accurate. Listen to specific instruments or vocalists, then to the overall mix. Be aware that to the human ear, louder always sounds better. If you aren't careful, all your sliders will be set to the top of their range! Your goal should be to keep most of the sliders at (or very near) their center detent positions.

Once you are fairly satisfied with your settings, try another cassette or CD and permit yourself minor touchups. Eventually you will find settings that sound right for most material. Use a marker to make a dot on the face of the equalizer next to each slider to show the preferred position. Now you (or your passengers) can make temporary changes without having to worry about setting the sliders back.

From now on, you may want to use the bass and treble controls to compensate for specific tapes or songs that need an extra dose of bass or treble. The broad characteristics of the tone controls are well suited for this type of compensation.

What Is Road Noise Masking?

First of all, what is road noise? *Road noise* refers to all the sounds associated with driving your car—the roar of your engine, the sound of your tires rolling on the pavement, and wind noise. Of course, road noise depends on the vehicle you're driving, how fast you're going, whether the windows are rolled up, and so forth. The important point is that, in general, road noise is dominated by deep bass frequencies.

Road noise masking means that you can't hear the deep bass of your music until it is loud enough to overcome the road noise. This isn't just interference—your ears actually reduce their sensitivity to deep bass in response to road noise! This means that to provide correct-sounding bass for normal listening levels, you need to boost your deep bass a bit. About 5 dB in the 45- to 100-Hz region is a good start, but let your ears have a vote.

Setting by RTA

You would think that having an RTA would make setting an equalizer a snap. Unfortunately, that's not the case. Using an RTA properly requires skill and patience.

The reason for this becomes apparent once you try to use one. You hook up the pink noise generator, turn on the stereo, and hold the calibrated test microphone about where your head would be. Now you move the microphone about 6 inches to the left and notice that the frequency response curve now looks nothing like the previous one. What happened?

What Is a Real-Time Analyzer?

A real-time analyzer provides a graphic display of the frequency response of a stereo system. Its calibrated pink noise* output is connected to the equalizer input of the system under evaluation, and then a calibrated microphone is used to monitor the sound output of the system. It's an invaluable tool for setting equalizers and for general frequency response evaluation and improvement.

The RTA of choice for car stereo work is the AudioControl SA-3055 (see Fig. 7-4; suggested retail price for the basic version is $1599). The SA-3055 is an upgraded version of the industry standard SA-3050. It's a $1/3$-octave real time analyzer with calibrated mike and printer interface. The term $1/3$ *octave* means the audio spectrum is broken into thirty bands. This resolution is necessary to display the narrowest peaks and valleys the human ear can readily perceive.

Audio analysis hardware/software packages for PCs often provide $1/3$-octave real-time analysis in addition to other useful tools such as speaker parameter measurement. Normally they include a module that plugs into your computer's printer port, software, and a calibrated microphone. Prices range from $700 to $5000, but look for budget units to start appearing.

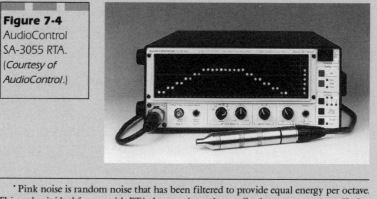

Figure 7-4
AudioControl
SA-3055 RTA.
*(Courtesy of
AudioControl.)*

---

*Pink noise is random noise that has been filtered to provide equal energy per octave. This makes it ideal for use with RTAs because it produces a flat frequency response display for flat frequency responses. It's called pink noise because it's generated from white noise with a filter that emphasizes low frequencies, and red corresponds to the low-frequency end of the visible light spectrum.

The problem is that direct and reflected sound add together to create a highly spatially dependent response. This means that you can't simply pick a spot to hold the microphone, then adjust the equalizer until you get a flat response.

One strategy is to use spatial averaging. This means that you use the average response from half a dozen strategically chosen microphone locations to set your equalizer. Easier said than done. Furthermore, this strategy glosses over the fact that the human ear is able to separate the direct response (the sound straight from the speakers) from the reflected response (the sound after it has reflected around inside your car). What sounds best is somewhere between a flat direct response and a flat total response. What the RTA measures is the total response.

The audio engineers at Ford, AudioControl, Pioneer, and other places have been trying for years to figure out how to automate setting an equalizer by using frequency response measurements from a microphone. One thing they agree on—it isn't simple.

The method I recommend for setting an equalizer with an RTA is explained below. It tries to achieve the desired blend between flat direct response and flat total response. It also relies on a tool that is impossible to automate—your ears! Note: This method is designed to provide optimized sound—it is not the method you would use if you were trying to win the ruler-flat frequency response portion of a sound-off competition.

Initial Preparation.   Start by setting your head unit bass and treble controls flat, and your equalizer sliders or knobs to their center detent positions. The fader control should be in the position you normally use when listening. Connect the pink noise output of the RTA to the inputs of the equalizer.

Presetting the Mid-Bass and Mid-Range.   Turn up the volume to a normal listening level. Set the balance control all the way to the left, so only the left speakers are playing. Position the microphone about 6 to 12 inches directly in front of the front left speaker, pointing right at it. This position strongly emphasizes the direct response over the reflected response. Now adjust the equalizer for flat response from 150 Hz to 1.5 kHz, trying to keep most of the sliders or knobs centered at their detent positions.

Setting the Deep Bass.   Position the mike in the driver's head position, pointing upward. Adjust the equalizer for flat response from 45 to 150 Hz. Leave the controls below 45 Hz set to their center detent positions. Boosting these ultra-low frequencies will not give you much benefit (few speaker systems can reproduce this range effectively)—it will instead dramatically reduce your amp headroom and cause your woofers to distort on those passages that have energy in this frequency range. To overcome the masking effects of road noise, you should now boost the response in the 45- to 100-Hz region by 5 dB or so.

Setting the Mid-Range and Treble.   Position the mike in the driver's head position, pointing toward the front left speaker. Adjust the equalizer for flat response from 1.5 kHz on up. Now move the microphone a foot to the right. Adjust the equalizer (1.5 kHz and up) to find a good compromise setting between the two positions. Try more mike positions in the vicinity of the driver's head to improve the blend further. In general, ignore frequency response valleys that do not occur for all mike positions. If you see a consistent strong peak or valley below 1.5 kHz, you should correct it as well, but for the most part, try to leave the lower frequency settings alone.

Touchup By Ear.   Return the balance control to the center detent position. Reconnect the equalizer to the head unit. Pick a CD or tape that you like and that sounds good on your home stereo. Choose one that covers the entire audio spectrum—something with bass drums, voice, and cymbals. Acoustic instruments such as piano, sax, and so on are better than synthesized ones because you know what they should sound like. Play your selection at a normal listening level.

Touch up the high-frequency (5 kHz and up) equalizer settings by ear. Maintain the general shape established previously, but add a gentle downward ramp as your ear dictates. It's normal to reduce the response at 16 kHz by 6 dB.

Look over the settings you have made. Any control above 150 Hz with more than 6 dB of boost compared to its neighbors should be checked by ear. Try reducing it by 3 dB and listening to decide which sounds better. Similarly, any control above 150 Hz with more than 10 dB of attenuation compared to its neighbors should be checked. (Response peaks sound worse than valleys, so the criteria are different.)

Use a marker to make a dot on the face of the equalizer next to each slider or knob to show the preferred position. Now you can make temporary changes without having to worry about setting the controls back right.

From now on, you will want to use the bass and treble controls to compensate for specific tapes or songs that need an extra dose of bass or treble. The broad characteristics of the tone controls are well suited for this type of compensation.

CHAPTER 8

Biamping and Crossovers

Biamping refers to using two amplifiers per channel (rather than the usual one amp per channel) to separately drive woofers and tweeters. Biamping offers an improvement in sound quality as well as increased tweeter protection, but it can be expensive. Biamping makes the most sense for high-power (50 watts per channel and above), high-performance systems.

More than almost any other subject, biamping requires knowledge of many areas of car stereo as well as subjects unique to itself. Rather than repeat information found in other chapters, I would like to direct your attention to them. Be sure to read Chap. 6 for important information on choosing and using amps. Chapters 3 and 4 contain similarly useful information on speakers and subwoofers, respectively.

Crossover Basics

To understand what biamping is all about, you need to understand the basics of crossovers.

The job of a crossover is to split the music signal into specific frequency bands—for example, low frequencies that go to a woofer, and high frequencies that go to a tweeter. Separate drivers are necessary because it's impractical to make a single driver that can cover the full audio range. Woofers can't effectively reproduce treble because of their large cone mass and high voice coil inductance. Tweeters can't effectively reproduce bass because their radiating surfaces are too small and their lightweight voice coils can't handle much power.

Speaker-Level and Preamp-Level Crossovers

For coax speakers, the crossover network is usually just a single capacitor in series with the tweeter, mounted on the speaker (Fig. 8-1).

Figure 8-1
Single-capacitor
crossover.

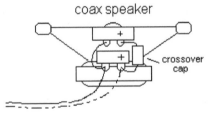

coax speaker

crossover
cap

The capacitor prevents low frequencies from going to the tweeter. There is no woofer filter—this low-budget approach relies on the natural high-frequency roll-off of the woofer. For separates, a better crossover (with an actual woofer low-pass filter) is usually provided (Fig. 8-2).

In both of these examples, the crossover is effectively *after* the amp, as represented in Fig. 8-3.

Crossovers after an amp are called *speaker-level crossovers* because the inputs and outputs are at speaker level. (Speaker-level crossovers are often called *passive crossovers* because they require no power, but this can be misleading since preamp-level crossovers can be passive too.) Speaker-level crossovers must handle high power levels and are built using large inductors and capacitors (Fig. 8-4).

Biamping instead puts the crossover *before* the amps, as shown in Fig. 8-5.

Figure 8-2
External crossover module.

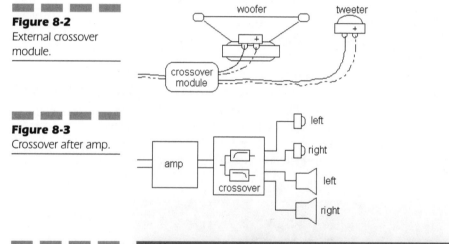

Figure 8-3
Crossover after amp.

Figure 8-4
Speaker-level crossover. (*Courtesy of MCM Electronics.*)

Figure 8-5
Crossover before the amps.

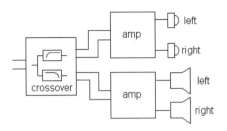

This is called biamping because it requires two amplifiers per channel instead of one. This type of crossover is called a *preamp-level crossover* because the input and output signals are at preamp level (Fig. 8-6).

Crossover Slopes

One of the most important characteristics of a crossover filter is the slope. *Slope* refers to how effectively the crossover filters out the unwanted frequencies.

Figure 8-7 shows the frequency response characteristics of four filters, each a 5-kHz high-pass filter. This type of filter would be used for a tweeter. The only difference between the filters is the slope—the shallowest slope is for a first-order (6 dB/octave) filter and the steepest slope is for a fourth-order (24 dB/octave) filter. Second-order (12 dB/octave) and third-order (18 dB/octave) filter responses are in between.

Figure 8-6
Electronic crossover.
(*Courtesy of Coustic.*)

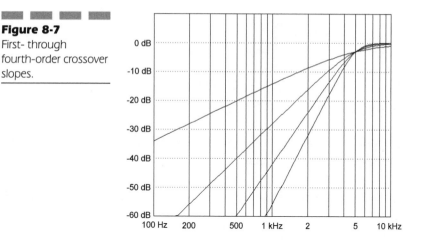

Figure 8-7

First- through fourth-order crossover slopes.

Above 5 kHz, all four filters are almost the same—they pass the signal with very little attenuation. This is where the similarity ends. At 2 kHz, the first-order filter is –8 dB. To put that into perspective, –10 dB sounds like half volume. This means that the tweeter is still strongly contributing to the total sound of the system at this frequency. More importantly, it's being subjected to lots of additional power. The fourth-order filter is –32 dB at 2 kHz. Not only will you not hear the tweeter at this level, it won't be subjected to dangerous power levels.

Why do you care if the tweeter is strongly contributing to the total sound below the crossover frequency? Two reasons. One, the frequency response of most tweeters starts to roll off and get bumpy below 2 kHz. This causes the total frequency response of the system to get bumpy as well. Two, where both the woofer and tweeter are trying to reproduce the same frequencies, an interference pattern is created. This causes spatially dependent reinforcement and cancellation of the sound, which degrades imaging. It's beneficial to reduce the range of frequencies where this happens by using sharper filter slopes.

The biggest concern about shallow slopes for tweeters is power handling. A tweeter that is rated at 50 watts (read the fine print: with a 5-kHz, 12-dB/octave crossover) is really only able to handle about 2 watts. This is because most of the power in music is in the bass—very little is in the treble. Table 8-1 shows the normal frequency distribution of power in music.

Using a shallow 6-dB/octave slope means huge amounts of additional mid-range and even bass power are directed to the tweeter. You can get away with this for conservatively powered systems (18 watts per channel or less) by using a high crossover frequency—for example, 8.5 kHz. Unfor-

TABLE 8-1

Normal Frequency
Distribution of
Power in Music

Frequency	Maximum Power Above That Frequency
300 Hz	50%
600 Hz	25%
1200 Hz	10%
2400 Hz	5%

tunately, this tactic usually results in a weak and bumpy system response near 5 kHz due to the high-frequency limitations of the woofer.

The bottom line is to avoid 6-dB/octave crossovers for the tweeter.

Why Biamp?

Everything just said about crossover slopes applies equally to both speaker-level and preamp-level crossovers. The problems with a shallow slope are exactly the same for speaker-level and preamp-level crossovers. You can get the same improvement by using a higher-order crossover, regardless of whether it's speaker level or preamp level.

So why would you want to put the crossover before the amps if you have to buy two amps instead of one? Good question. There are three main advantages to using preamp-level crossovers:

- Increased practicality of high-order slopes
- Driver impedance characteristic doesn't affect frequency response
- Tweeter protection during clipping of bass

Combined, these factors provide an improvement in sound quality as well as increased tweeter protection under a number of circumstances.

Increased Practicality of High-Order Slopes

Speaker-level crossovers must handle the high power levels of the amplifier outputs. This means massive high-power inductors and capacitors are required. The steeper the crossover slope, the more components required. In practice, anything higher than 12 dB/octave is considered cost prohibitive.

Preamp-level crossovers are low-power circuits and eliminate the need for expensive high-power components. With active filters, the cost of implementing a 24-dB/octave slope is not much more than that for a 6-dB/octave slope. Active filters make it simple to provide switch-selectable crossover frequencies. They are also immune to many of the performance problems, caused by inductor resistance and saturation and electrolytic capacitor degradation due to aging, that plague speaker-level crossovers. In short, they act more like ideal filters.

Driver Impedance Characteristic Doesn't Affect Frequency Response

The impedances of woofers and tweeters vary dramatically with frequency. This makes it difficult for speaker-level crossovers to work as intended. The result is frequency response anomalies.

So-called universal speaker-level crossovers that you can buy assume that your speakers are 4 ohms for all frequencies. Crossovers that are custom designed for a specific woofer and tweeter offer some improvement, but there are practical limitations.

Biamping solves this problem because the amplifiers directly drive the speakers. Since the output impedance of the amplifiers is small compared to that of the drivers, driver impedance variations with frequency have no effect on the response.

Tweeter Protection During Clipping of Bass

In a single-amp system, when you turn up the volume of your stereo loud enough to cause distortion, massive amounts of high-frequency energy are produced at the output of the amplifier. This is because distorting amplifiers "clip" the musical peaks. This generates high-frequency harmonics not in the original signal. Tweeters are often unable to handle the additional power, and are damaged or destroyed.

Ironically, using an underpowered amp (in a normal speaker-level crossover situation) risks blowing tweeters more than using an overpowered amp, because of clipping.

Most clipping occurs during loud bass passages, such as drumbeats. Even though a loud drumbeat may only contain bass frequencies, high-power "spikes" are produced at high frequencies when the amp clips. These spikes get to the tweeters with a single-amp system, but not with biamping. Biamping eliminates this problem because the tweeters are powered by their own amp, which is unaffected by the woofer amp. Clipping of the tweeter amp is not a concern because the total power to the tweeters is not increased.

What to Look For in Equipment

Before you start shopping for equipment, study the project sections ("Biamping Head Units with Two Sets of Preamp Outputs," "Biamping Head Units with One Set of Preamp Outputs") to see which configuration best suits your needs. If you're buying a head unit, get one with two sets of preamp outputs so you can be sure to have full use of your fader.

Choosing a Crossover

There are a number of considerations for choosing the best crossover for your application. These include crossover configuration, filter slopes, and choice of crossover frequencies.

A common configuration is shown in Fig. 8-8. All inputs and outputs are preamp level.

There is no standard configuration for crossovers, but this configuration is assumed for the projects section because it's the most prevalent. Other variations exist, and may be better suited to your needs. For example, speaker-level inputs would be important if your head unit doesn't have preamp-level outputs. Manufacturers' "front" and "rear" designations simply indicate which inputs are used for each of the outputs—they may not be accurate for your application. Similarly, manufacturers often refer to the woofer outputs as the mid-band or bandpass outputs, and the subwoofer outputs as low-pass. It is standard for all output pairs to have level controls.

The requirements for crossover filter slopes and frequencies are shown in Table 8-2.

Figure 8-8
Common crossover configuration.

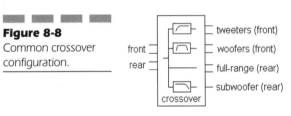

front
rear

tweeters (front)
woofers (front)
full-range (rear)
subwoofer (rear)

crossover

TABLE 8-2

Crossover
Requirements

Crossover Characteristic	Subwoofer Low-Pass	Woofer High-Pass	Woofer Low-Pass	Tweeter High-Pass
Filter slopes	18 dB/octave minimum	6 dB/octave minimum	12 dB/octave minimum	12 dB/octave minimum
	Need not be the same as woofer high-pass slope	Need not be the same as subwoofer low-pass slope	Should be the same as tweeter high-pass slope	Should be the same as woofer low-pass slope
Crossover frequencies	75 to 150 Hz minimum range	100 to 200 Hz minimum range	2.4 to 5 kHz minimum range	2.4 to 5 kHz minimum range
Other important features	40 or 45 Hz bass boost			
	Subsonic filter			
	Polarity switch			

The tweeter crossover slope must be 12 dB/octave minimum for tweeter power handling considerations. Anything less forces you to use a crossover frequency that is too high for good woofer response. The woofer low-pass slope and crossover frequency should match those of the tweeter for best performance.

To enable placement of a subwoofer anywhere in the vehicle, the subwoofer crossover must be 18 dB/octave or higher. Slopes less than 18 dB/octave allow mid-bass and even mid-range to be heard from the subwoofer. Since only deep bass frequencies are nondirectional, this gives away the subwoofer location and degrades imaging. Who wants a bass guitar player under their seat? The only exception to this is for bandpass boxes, where you can successfully use a 12-dB/octave crossover. Either buy a crossover that has a steep enough subwoofer slope or use a separate subwoofer crossover that does.

A continuously variable subwoofer cutoff frequency is a real plus. It gives you the flexibility to seamlessly blend the subwoofer with the rest of your system. Even with high-pass crossovers on your woofers, the natural low-frequency roll-off of woofers makes it difficult to predict the best subwoofer crossover frequency. A continuously variable crossover lets you avoid a gap or peak in your mid-bass.

Five to 10 dB of bass boost at 40 or 45 Hz extends the low-frequency response of most subwoofer systems without overdriving them. A subsonic filter and polarity switch are other important features for a subwoofer crossover. See Chap. 4 for a detailed explanation.

Buying Amps for Biamping

Most of what you should know about buying amps is contained in Chap. 6. For biamping, you also need to know the relative power levels for tweeter amps, woofer amps, and subwoofer amps (if you have them).

This comes down to the normal frequency distribution of power in music. Table 8-3 shows the maximum percentage of power above various frequencies.

How much power a tweeter amp should have depends on the crossover frequency. The higher the crossover frequency, the less the required power. Rarely is a tweeter crossover frequency lower than 2400 Hz, so a tweeter amp having roughly 5 percent of the total system power should be sufficient.

On the subwoofer end, the power requirements also depend on the crossover frequency. The lower the crossover frequency, the less the required power for the sub amp, but the more the required power for the woofer amp.

The rules of thumb in Table 8-4 should help you determine the right power levels for tweeter and subwoofer amps, based on the total system power. The tweeter recommendation applies whether you have a subwoofer or not.

For example, suppose you're biamping in front and back and don't plan to use a subwoofer. If you're using 100-watt per channel woofer amps, then your tweeter amps should have at least 5 watts per channel. (In prac-

TABLE 8-3

Normal Frequency
Distribution of
Power in Music

Frequency	Maximum Power Above That Frequency
300 Hz	50%
600 Hz	25%
1200 Hz	10%
2400 Hz	5%

TABLE 8-4

Amplifier Power
Requirements for
Biamping

Amps	Power
Tweeter amps	5% of total system power
Sub amps	40% of total system power

tice, you'll have to buy 18-watt per channel amps, since they're the smallest available.)

Suppose you plan to have a subwoofer as well as biamp in front and back. Let's say you're shooting for a total system power of around 400 watts (about the same as the last example). Once again, the tweeter amps should have at least 5 watts per channel. (You'll have to buy 18-watt per channel amps.) The total power to the subwoofer should be about 160 watts. The total power to the woofers would be 220 watts or 55 watts per channel.

Speakers

For biamping, you'll either need to use separate woofers and tweeters (component speakers) or biampable coax drivers. Biampable coax drivers are like regular coaxial speakers, but they provide separate terminals for the woofer and tweeter. These save you from having to mount separate tweeters, but you lose the flexibility of being able to put the tweeters wherever you want.

More about selecting and mounting component speakers is contained in Chap. 3. Subwoofers are covered in Chap. 4.

Biamping Head Units with Two Sets of Preamp Outputs

Having two sets of preamp outputs is a big plus for biamping because it lets you retain full use of your head unit fader in most configurations. There are five biamping configurations for head units with two sets of preamp outputs (Table 8-5)—all but the second and third retain use of the fader. The first configuration provides the most flexibility and highest performance, but the fourth and fifth provide the best performance value.

TABLE 8-5

Biamping
Configurations for
Head Units with
Two Sets of Preamp
Outputs

Configuration	Diagram	Comments
Separate rear crossover	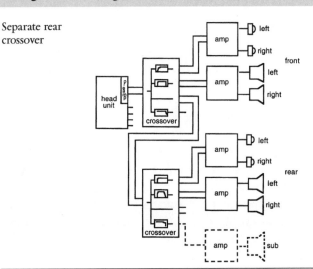	Allows you to retain full use of your head unit fader. Lets you choose different crossover frequencies for the front and rear. Crossovers may be purchased as a single combined unit.
Shared crossover	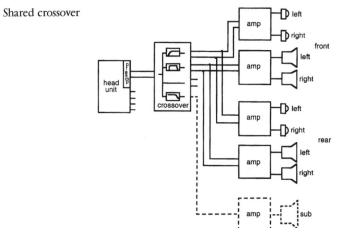	Head unit fader becomes non-functional. Difficult to make front/rear level adjustments since you must also maintain the proper relative tweeter/woofer levels. Not recommended.

TABLE 8-5

(continued)

Configuration	Diagram	Comments
Parallel front/rear speakers		Saves two amps when equal power to front and rear speakers is acceptable. (No relative front/rear level adjustment is possible.)
Full-range rear		Allows you to retain full use of your head unit fader. Saves the cost of an amplifier, and you only need a single crossover. Provides virtually same performance as separate rear crossover configuration.
Amp-saving full-range rear		Allows you to retain full use of your head unit fader. Useful for rear fill applications where high power is needed for front only. Saves the cost of two amplifiers, and you only need a single crossover.

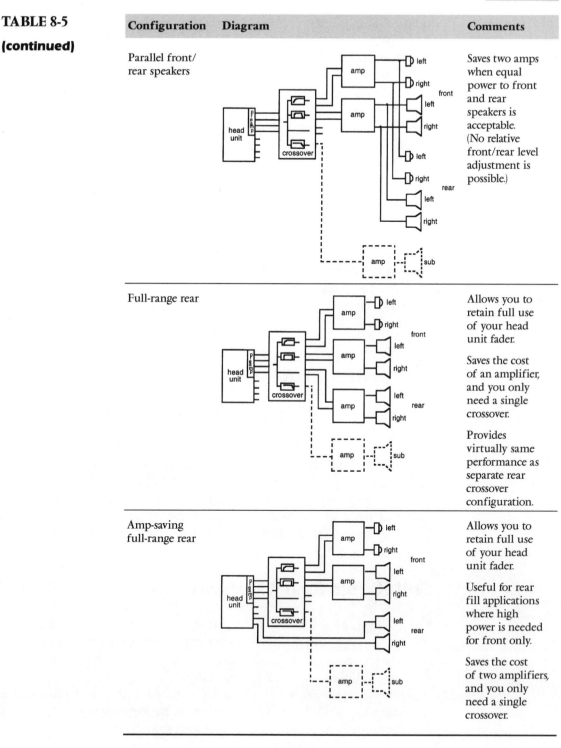

Separate Rear Crossover Configuration

This configuration provides the most flexibility and highest performance, but is the most expensive. It uses separate crossovers for the front and rear speakers, which allows you to retain full use of your head unit fader. This also lets you choose different crossover frequencies for the front and rear (of limited value). The crossovers may be purchased as two separate units or as a single combined unit.

Shared Crossover Configuration

This configuration shares a single crossover for front and rear component speakers. The main problem is the loss of head unit fader functionality—you must use the amp level controls to set the relative front/rear levels. Since you must also maintain the proper relative tweeter/woofer levels, it becomes a chore to make slight front/rear readjustments. For this reason, this configuration is not recommended.

Parallel Front/Rear Speakers Configuration

The parallel front/rear speakers configuration shares amps for front and rear channels. This gives you almost the same power per speaker as having identical sets of amps for front and rear. The drawback with this configuration is that you are stuck with having equal power for front and back speakers.

Since the amplifiers used in this configuration will see a parallel speaker load, they must be 2-ohm stable. Amps driving 2-ohm loads run hotter, so be sure to mount them where they will have good air circulation.

Full-Range Rear Configuration

This configuration biamps only the front set of speakers—the rear speakers are full-range (usually coaxial) speakers. This saves the cost of an amplifier and you only need a single crossover. This configuration allows you to retain full use of your head unit fader.

Since the performance of the front speakers is the most critical, this configuration provides virtually the same performance as the separate rear crossover configuration, but at a much lower cost.

Amp-Saving Full-Range Rear Configuration

This configuration is identical to the previous one, except it saves an additional amplifier by using the head unit's internal amp to drive the rear speakers. This makes sense if you are using the rear speakers for rear fill and operating them at a reduced volume level to improve imaging (see the "Imaging Overhaul" section in Chap. 3). This configuration allows you to retain full use of your head unit fader as well.

In general, providing rear fill requires one-tenth the power necessary for the front. Assuming a 10-watt per channel head unit, this configuration should work for up to 100 watts per channel in the front.

Biamping Head Units with One Set of Preamp Outputs

Having only one set of preamp outputs is a minus for biamping because you generally lose use of your head unit fader. There are five biamping configurations for head units with one set of preamp outputs (Table 8-6)—only the fifth retains use of the fader. The first configuration provides the most flexibility and highest performance otherwise, but the fourth and fifth provide the best performance value.

Separate Rear Crossover Configuration

This configuration provides the most flexibility (except for the loss of fader) and the highest performance. It is the most expensive as well. It uses separate crossovers for the front and rear speakers. This lets you choose different crossover frequencies for the front and rear (of limited value). Although the head unit fader control is nonfunctional, setting the relative front/rear level is facilitated by having the level of all rear speakers affected by a single control (the full range output control of the front crossover). The crossovers may be purchased as two separate units or as a single combined unit.

Shared Crossover Configuration

This configuration shares a single crossover for front and rear component speakers. The head unit fader control is nonfunctional. Setting the relative front/rear levels is especially difficult because there is no level control

TABLE 8-6

Biamping
Configurations for
Head Units with
One Set of Preamp
Outputs

Configuration	Diagram	Comments
Separate rear crossover	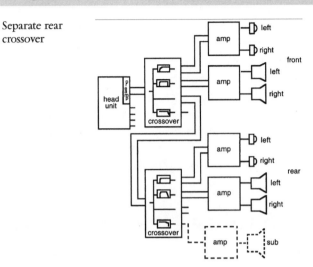	Head unit fader control is non-functional, but setting the relative front/rear level is facilitated by having a common rear level control. Lets you choose different crossover frequencies for the front and rear. Crossovers may be purchased as a single combined unit.
Shared crossover		Head unit fader is non-functional. Difficult to make front/rear level adjustments since you must also maintain the proper relative tweeter/woofer levels. Not recommended.

TABLE 8-6

(continued)

Configuration	Diagram	Comments
Parallel front/ rear speakers	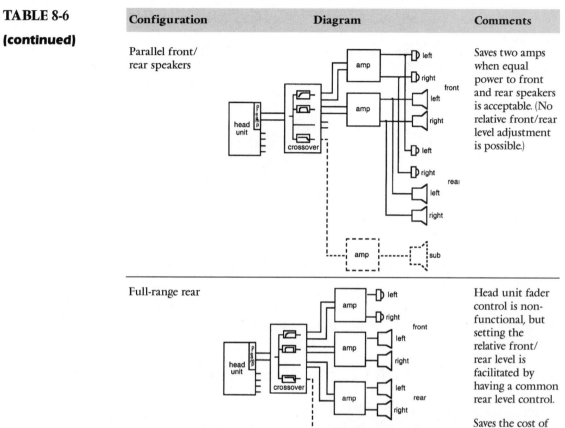	Saves two amps when equal power to front and rear speakers is acceptable. (No relative front/rear level adjustment is possible.)
Full-range rear		Head unit fader control is non-functional, but setting the relative front/rear level is facilitated by having a common rear level control.

Saves the cost of an amplifier, and you only need a single crossover.

Provides virtually the same performance as the separate rear crossover configuration. |

TABLE 8-6

Biamping
Configurations for
Head Units with
One Set of Preamp
Outputs
(continued)

Configuration	Diagram	Comments
Amp-saving full-range rear		Allows you to retain full use of your head unit fader. Useful for rear fill applications where high power is needed for front only. Saves the cost of two amplifiers, and you only need a single crossover.

that affects only all front or only all rear speakers. For this reason, this configuration is not recommended.

Parallel Front/Rear Speakers Configuration

The parallel front/rear speakers configuration shares amps for front and rear channels. This gives you almost the same power per speaker as having identical sets of amps for front and rear. The drawback with this configuration is that you are stuck with having equal power for front and back speakers.

Since the amplifiers used in this configuration will see a parallel speaker load, they must be 2-ohm stable. Amps driving 2-ohm loads run hotter, so be sure to mount them where they will have good air circulation.

Full-Range Rear Configuration

This configuration biamps only the front set of speakers—the rear speakers are full-range (usually coaxial) speakers. This saves the cost of an amplifier, and you only need a single crossover. Although the head unit fader control is nonfunctional, setting the relative front/rear level is facilitated by having the level of all rear speakers affected by a single amplifier control.

Since the performance of the front speakers is the most critical, this configuration provides virtually the same performance as the separate rear crossover configuration, but at a much lower cost.

Amp-Saving Full-Range Rear Configuration

This is the only configuration that provides head unit fader control for head units with one set of preamp outputs. Like the previous configuration, this one biamps only the front set of speakers, but it saves an additional amplifier by using the head unit's internal amp to drive the rear speakers. This makes sense if you are using the rear speakers for rear fill and operating them at a reduced volume level to improve imaging (see the "Imaging Overhaul" section in Chap. 3).

In general, providing rear fill requires one-tenth the power necessary for the front. Assuming a 10-watt per channel head unit, this configuration should work for up to 100 watts per channel in the front.

Adjusting Your System

Once you've installed everything, it's time to set all the crossover frequencies and amp levels. Follow the order below for best results.

Setting Tweeter High-Pass and Woofer Low-Pass Crossover Frequencies

Setting the tweeter high-pass and woofer low-pass crossover frequencies should be done prior to the other system adjustments.

The optimum tweeter high-pass and woofer low-pass crossover frequency settings depend on the specific woofer and tweeter involved. If you are fortunate enough to have recommended crossover frequencies or frequency response data for your drivers, you can base your crossover settings on that information. Otherwise, you'll have to make an educated guess.

■ *If your component speakers came with speaker-level crossovers:* Component speakers bought as a matched set often come with speaker-level crossovers. Using the same crossover frequencies as the speaker-level crossovers is your best bet.

- *If you have frequency response curves for your speakers:* Ideally, the high-frequency roll-off point of the woofer extends above the low-frequency roll-off point of the tweeter. If so, the woofer and tweeter crossover frequencies should both be set to the same frequency—an equal compromise between woofer and tweeter roll-off points. If the high-frequency roll-off point of the woofer does not extend above the low-frequency roll-off point of the tweeter, you will need to set the woofer crossover frequency higher than the tweeter crossover frequency. This usually means using 5 kHz for the woofer and 3.5 kHz for the tweeter.

- *If you don't have anything (except the chart in Table 8-7):* In general, the smaller the woofer, the more extended its high-frequency response. Table 8-7 is based on that fact.

Regardless of which method you use, don't be afraid to use your ears and experiment with other settings later, after you've made a first pass at all the other system settings.

Setting Woofer High-Pass Crossover Frequencies

If you don't have a subwoofer, and you're using the rear speakers only for fill, then you should not use any woofer high-pass filtering at all. If you don't have a subwoofer, and you play your rear speakers loud enough to provide bass for the whole system, then consider using 150-Hz high-pass filtering for the front woofers. This will let them play louder with less distortion, and will not significantly reduce the total bass of the system. This is especially true for small front woofers.

If you do have a subwoofer, then you should use high-pass filtering for all your woofers. Use 200 Hz for 6-dB/octave woofer high-pass slopes and 150 Hz for 12-dB/octave or higher slopes.

TABLE 8-7

Crossover Frequency Settings Based on Woofer Diameter

Woofer Diameter	Woofer Crossover Frequency	Tweeter Crossover Frequency
4″	3.5 kHz	3.5 kHz
5¼″	3.5 kHz	3.5 kHz
6½″	5 kHz	3.5 kHz

Setting Woofer and Tweeter Amp Levels

Setting woofer and tweeter amp levels is a complicated process. You need to simultaneously set the proper relative tweeter-to-woofer levels, the proper relative front-to-rear levels, and the optimum absolute levels for lowest noise and distortion. On top of that, you probably have level controls on the crossover outputs as well as amplifier inputs to worry about.

Follow the procedure for your specific configuration.

Separate Rear Crossover and Shared Crossover Configurations.

1. Set the head unit fader control to its center detent position. Set all tone controls and equalizer controls flat.

2. Set the level controls of all amps to minimum. Set all crossover level controls to maximum.

3. With music, turn up the head unit volume control (you should barely be able to hear the music). When you reach the point where you start to hear distortion, turn the volume control down slightly until the distortion disappears. Leave the volume control at this setting.

4. Turn up the front woofer amp level control until you the system is as loud as you'll ever play it or you begin to hear distortion. If you hear distortion, slightly decrease the amp level control.

5. Turn down the head unit volume to a normal listening level.

6. Turn up the rear woofer amp level until the desired front/rear balance is achieved.

7. With the fader set all the way to the front, turn up the front tweeter amp to achieve the proper tonal balance.

8. With the fader set all the way to the rear, turn up the rear tweeter amp to achieve the proper tonal balance. Return the fader to its center detent position.

Parallel Front/Rear Speakers Configurations.

1. Set the head unit fader control to its center detent position. Set all tone controls and equalizer controls flat.

2. Set the level controls of all amps to minimum. Set all crossover level controls to maximum.

3. With music, turn up the head unit volume control (you should barely be able to hear the music). When you reach the point where you start to hear distortion, turn the volume control down slightly until the distortion disappears. Leave the volume control at this setting.

4. Turn up the woofer amp level control until you the system is as loud as you'll ever play it or you begin to hear distortion. If you hear distortion, slightly decrease the amp level control.

5. Turn down the head unit volume to a normal listening level.

6. Turn up the tweeter amp to achieve the proper tonal balance.

Full-Range Rear (Not Amp-Saving) Configurations.

1. Set the head unit fader control to its center detent position. Set all tone controls and equalizer controls flat.

2. Set the level controls of all amps to minimum. Set all crossover level controls to maximum.

3. With music, turn up the head unit volume control (you should barely be able to hear the music). When you reach the point where you start to hear distortion, turn the volume control down slightly until the distortion disappears. Leave the volume control at this setting.

4. Turn up the (front) woofer amp level control until you the system is as loud as you'll ever play it or you begin to hear distortion. If you hear distortion, slightly decrease the amp level control.

5. Turn down the head unit volume to a normal listening level.

6. Turn up the (front) tweeter amp to achieve the proper tonal balance.

7. Turn up the (rear) full-range amp level until the desired front/rear balance is achieved.

Amp-Saving Full-Range Rear Configurations.

1. Set the head unit fader control to its center detent position. Set all tone controls and equalizer controls flat.

2. Set the level controls of all amps to minimum. Set all crossover level controls to maximum.

3. With music, set the head unit volume to a normal listening level.

4. Turn up the (front) woofer amp level control until the desired front/rear balance is achieved.

5. Turn up the (front) tweeter amp to achieve the proper tonal balance.

Setting Subwoofer Crossover Frequency and Amp Level

Perform the adjustments in this order for best results. Use your favorite cassette, CD, or radio station. Use several sources to get a good average.

Preset the subwoofer crossover to 100 Hz. If there is a 40- or 45-Hz bass boost feature, turn it off for now. Adjust the subwoofer amp level control to the point that sounds best, erring on the side of too little bass. If there is a subwoofer polarity switch, set it to whichever position gives you the most bass. The effect is usually small.

If you plan to use the 40- or 45-Hz bass boost feature, turn it on now. In general, it will help extend the system bass response and provide tighter-sounding bass. If you later find that your subwoofer sometimes distorts, you should turn the bass boost off.

Now fine-tune the cutoff frequency of the subwoofer crossover and the level control of the subwoofer power amplifier to provide the most natural-sounding bass. You will find that you need more bass when you are driving to overcome the masking effects of road noise. (See the box on road noise masking on p. 196 in Chap. 7.) Try to find a level that is acceptable for all conditions.

CHAPTER **9**

CD Changer
Projects

Adding a CD changer is a great way to add CD capabilities to your system without having to replace your head unit. If you listen to CDs *and* cassettes, a CD changer lets you have both. And there's nothing better than a changer loaded with your favorite discs for long road trips.

Adding a CD Changer to Your System

Unlike most other car stereo projects, adding a CD changer strongly depends on the type of head unit you have. That's because many head units don't have CD changer controls, and there is no industry standard for those that do. It's important to scope out the interfacing issues before you start choosing features or brands of equipment.

For the purposes of adding an aftermarket CD changer, there are three head unit categories: head units without CD changer controls, aftermarket head units with CD changer controls, and factory head units with CD changer controls. Table 9-1 provides an overview of these head unit categories.

Head Units Without CD Changer Controls

Most older head units (and many new ones) have no provisions for adding a CD changer. Many people deal with this situation by replacing their head unit with one having CD changer controls. If you want to keep the

TABLE 9-1

Head Unit Categories for Adding a CD Changer

Head Unit	Comments
Head unit without CD changer controls	An aftermarket CD changer can be added using an FM modulator or CD changer with amplifier.
Aftermarket head unit with CD changer controls	No performance compromises and all your controls are in one place.
	You are generally restricted to choosing the same brand and vintage of equipment for both changer and head unit.
Factory head unit with CD changer controls	Adding an aftermarket CD changer is often possible using an OEM interface CD changer converter.

head unit you have, the aftermarket industry offers a number of solutions (Table 9-2).

CD Changer with FM Modulator.   The most common solution to the problem of a head unit lacking CD changer controls is the FM modulator approach (Fig. 9-1).

With this approach, the CD changer effectively becomes a personal radio station for your FM receiver. The FM modulator converts the audio signals of the CD changer to an FM stereo signal that is injected into your existing antenna wire. This assures good reception. You can choose the frequency of your personal radio station to avoid a conflict with a commercial station.

When you want to listen to the CD changer, you simply tune your FM receiver to your "personal station." Your head unit controls (such as volume, bass, and treble) function normally. The CD changer itself is controlled through a dash-mounted display panel with controls. A wireless remote control is offered with some models.

The weakness of the FM modulator approach is the degradation in the signal quality due to the modulation/demodulation process. Frequency response, signal-to-noise ratio, distortion, and stereo separation all suffer. Because of this, purists often consider the FM modulator approach unacceptable. But to put this in perspective, consider the sound quality of the most noise-free, best-sounding FM station you've ever heard. That's what

Tip: If you travel, choose a model that lets you change the transmitting frequency from the control panel.

TABLE 9-2

Solutions for Head Units Without CD Changer Controls

Solution	Comments
CD changer with FM modulator	The CD changer becomes a "personal radio station" for your FM receiver.
	The head unit controls volume, bass, and treble for all sources. The CD changer is controlled through a dash-mounted display panel.
	Some degradation in the signal quality due to the modulation/demodulation process, although the audio quality can be much better than a normal FM broadcast.
CD changer with amplifier	Avoids any degradation in signal quality due to FM modulation.
	Can provide higher power to your speakers.
	The CD changer control panel controls volume, bass, and treble for all sources.

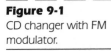

Figure 9-1
CD changer with FM
modulator.

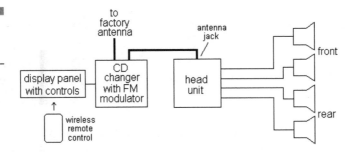

you'll get. In fact, because of the heavy compression used by commercial broadcasters to reach the maximum audience, the audio quality can be much better than any FM broadcast.

CD Changer with Amplifier. An innovative solution to the problem of head units lacking CD changer controls is the CD changer with amplifier (Fig. 9-2).

By integrating amplifiers and controls with the changer, it becomes possible to avoid any degradation in signal quality due to FM modulation and to provide higher power to your speakers. The JVC KD-MA1 (Fig. 9-3) uses this approach.

With the FM modulator approach, the CD changer is a signal source for the head unit. The head unit remains the control center for everything but CD changer functions. With the CD changer with amplifier approach, the head unit becomes a signal source for the CD changer! The CD changer becomes the primary control center.

Figure 9-2
CD changer with
amplifier.

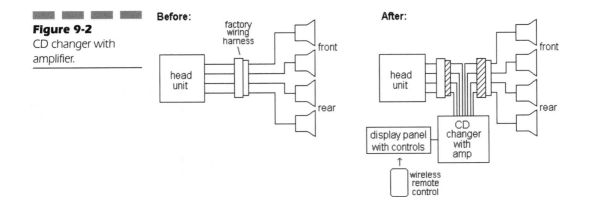

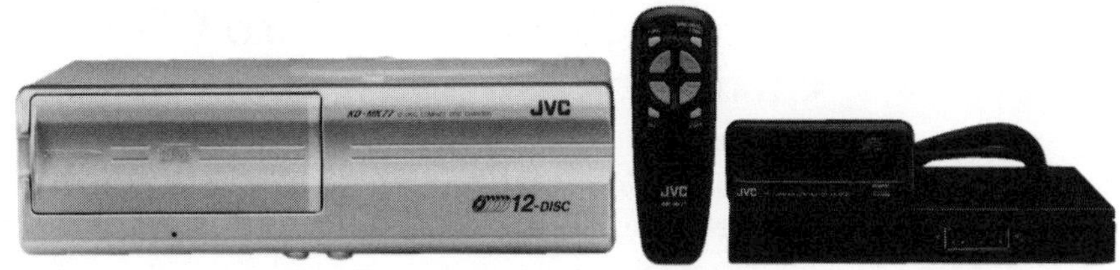

■■ ■■ ■■ ■■
Figure 9-3
JVC KD-MA1 CD
changer with amplifier.
(*Courtesy of JVC.*)

When you want to listen to the CD changer, all the controls (including volume, bass, and treble) are on the dash-mounted display panel. When you want to listen to the radio or in-dash cassette player, use the head unit controls only for tuning, rewinding, and so on—continue to use the dash-mounted display panel for volume, bass, and treble. A wireless remote control can also be used for control panel functions.

If your head unit lacks high-power outputs, then the changer with amplifier approach gives you another benefit besides sound quality—more power to drive your speakers. Direct plug-in harnesses are available with the JVC KD-MA1 to make wiring a breeze in many Ford, Honda, Nissan, Toyota, GM, and Chrysler vehicles.

The only drawback to the CD changer with amplifier approach is getting used to using the CD changer control panel for volume control of the radio and tape deck, and leaving the head unit control alone.

Which approach is better—FM modulator or changer with amplifier? If most of your listening is to the CD changer, then the changer with amplifier approach comes out ahead. If you use the CD changer less often, the FM modulator approach makes sense.

Aftermarket Head Units with CD Changer Controls

With an aftermarket head unit with CD changer controls, it's easy to add a CD changer. There are no performance compromises and all your controls are in one place. You can even control everything with a single remote.

It's important to realize that *not all CD changers work with all head units*. In fact, you are generally restricted to choosing the same brand and vintage of equipment for both changer and head unit. This means that you should look closely at CD changers when you're choosing a head unit. It also means that you shouldn't put off buying a CD changer too long after buying a head unit, or incompatibility may become a problem.

Factory Head Units with CD Changer Controls

Adding an aftermarket CD changer to a factory head unit with CD changer controls is often possible using an OEM interface CD changer converter.

OEM interface CD changer converters connect between the factory head unit and the aftermarket CD changer. For some applications, it's simply a matter of converting from one connector type to another. In other cases, the handshaking protocol between the head unit and changer must be digitally converted from one format to another.

Precision Interface Electronics (PIE) manufactures OEM interface CD changer converters for many vehicles (Fig. 9-4). Each of these is designed to work with a specific brand and vintage of CD changer, so don't buy a changer until you're sure a suitable converter exists.

Performance and Features

You can expect excellent frequency response and signal-to-noise ratio with virtually any CD changer. Features like intro scan, random play, and track repeat are found on almost all models. Look to the number of CDs and disc naming/track programming features to help you choose the right changer. If you go off road, electronic shock protection can help reduce skipping. Don't forget to consider size and mounting angle, especially if you're thinking about a glove compartment or console installation. Table 9-3 provides information on various CD changer features.

Number of CDs. Most changers hold six, ten, or twelve CDs. Some handle as many as fifty discs.

Figure 9-4
PIE HS-1A OEM interface CD changer converter. (*Courtesy of Precision Interface Electronics.*)

TABLE 9-3

Performance and
Features of CD
Changers

Performance/Feature	Comments
Number of CDs	Choose 6, 10, 12, or 50 discs.
Disc naming/track programming	Lets you program disc names to appear on the head unit's display and preprogram the songs you want to hear on each disc.
FM modulator	Important for head units lacking CD changer controls. The CD changer effectively becomes a personal radio station for your FM receiver.
Mounting angle	Defines the range of mounting angles for proper operation.
CD frequency response	The frequency range a CD player can faithfully reproduce.
Electronic shock protection	A buffer memory used as a reserve against skipping.
Intro scan	Lets you hear the first few seconds of each track. Hit the button again when you hear the song you want.
Random play/shuffle	Mixes up the order of songs for variety during playback.
CD signal-to-noise ratio	A measure of how well a CD player silences background noise.
Zero-bit detect	Mutes the output whenever a series of zeros is detected in the digital bitstream.

Disc Naming/Track Programming. Disc naming lets you program disc names to appear on the head unit's display. Track programming allows you to preprogram the songs you want to hear on each disc. Most changers with these features will store information for about 100 discs.

FM Modulator. Important for adding a CD changer to a head unit lacking CD changer controls. With this approach, the CD changer effectively becomes a personal radio station for your FM receiver. The FM modulator converts the audio signals of the CD changer to an FM stereo signal that is injected into your existing antenna wire. You can choose the frequency of your personal radio station to avoid a conflict with a commercial station.

Mounting Angle. Defines the range of mounting angles for proper operation. Almost all models allow 0 and 90° (90° means "face up"). Some allow mounting over the entire 0- to 90° range or more, important for glove compartment installations.

CD Frequency Response.   The frequency range a CD player can faithfully reproduce. Humans can hear sounds as low as 20 Hz and as high as 20 kHz. Virtually every CD player exceeds this range.

Electronic Shock Protection.   Electronic shock protection is a buffer memory used to store between 1 and 10 seconds of music as a reserve against skipping due to bumps in the road.

Intro Scan.   Lets you hear the first few seconds of each track. Hit the button again when you hear the song you want.

Random Play/Shuffle.   Mixes up the order of songs for variety during playback.

CD Signal-to-Noise Ratio.   A measure of how well a CD player silences background noise. Ratings are given in decibels; higher ratings indicate less noise. The signal-to-noise ratio of CD players is generally so good that noise picked up by other means determines the system noise level. Unless you're a competitor or plan on using super high-power amps, don't rely too heavily on this spec.

Zero-Bit Detect.   Mutes the output whenever a series of zeros is detected in the digital bitstream. The result is complete silence between songs. This is important if you're a competitor, but otherwise you'll never hear the difference.

Installation

Installing a CD changer is a matter of choosing a mounting location, mounting the changer, and connecting it up. It's a good idea to understand the basics of installation beforehand, to help you choose the right changer and so you know what to expect.

Choosing a Mounting Location

No single mounting location is best for every vehicle, changer, or person (Table 9-4). Think about where you might want to put your changer before shopping for one—it can determine the size and mounting angle

Tip: When choosing a spot to mount the changer, be sure you can easily remove the disc magazine. This is a common oversight for under-seat mounting.

range you'll need. Grab a tape measure and scope out your vehicle, then look at sizes of CD changers. The Crutchfield catalog has dimensions for every changer Crutchfield carries, so you can easily get an idea of what is possible.

Trunk.   In the trunk, finding space for a changer is rarely a problem. If you choose a fifty-disc changer, the trunk is probably your only option. In vehicles without a trunk release lever, the trunk provides the mounting location least vulnerable to theft. A disadvantage of trunk mounting is accessibility—you need to change disc magazines from outside the vehicle. If your trunk is packed with luggage, changing disc magazines is a chore. You also need to be careful about the changer getting wet if you open the trunk when it's raining.

   If you want to mount your CD changer at an angle (against the sloping front of the trunk, for example), make sure you choose a changer that allows mounting over the entire 0- to 90° range.

Under Seat.   The beauty of mounting a changer under a front seat is that it takes advantage of otherwise unused space. The amount of space under a seat varies quite a bit from vehicle to vehicle. When measuring for space, you need to check all the possible positions of the seat for interference, including front-to-back position, tilt, and height settings. A disadvantage of mounting a changer under the seat is its susceptibility to dirt, spills, and getting kicked from behind.

TABLE 9-4

CD Changer Mounting Locations

Mounting Location	Comments
Trunk	May be least vulnerable to theft.
	You need to change disc magazines from outside the vehicle. If your trunk is packed with luggage, changing disc magazines is a chore.
Under seat	Takes advantage of otherwise unused space.
	Susceptible to dirt, spills, and getting kicked from behind.
Console	Lets you comfortably change discs from the driver's seat.
Glove compartment	Most glove compartments are too small to hold even a six-disc changer.
	Usually requires an odd mounting angle, so choose a changer that allows mounting over the entire 0- to 90° range.

Console. Console mounting lets you comfortably change discs from the driver's seat. Space is tight, so you may need to choose a six-disc changer. The obvious disadvantage of this location is the loss of most of the useable console space.

Glove Compartment.   Most glove compartments are too small to hold even a six-disc changer. If your vehicle is one of the select few, glove compartment installation provides convenient access to the changer. Glove compartment installation usually requires an odd mounting angle, so choose a changer that allows mounting over the entire 0- to 90° range.

Mounting

Almost all CD changers can be mounted at 0 and 90° (90° means "face up"). Many allow mounting over the entire 0- to 90° range. You'll need to adjust the changer's suspension to match the mounting angle. Check the owner's manual for instructions.

Make sure the mounting surface you've chosen is strong enough to support the changer firmly, so it can't rip loose during a sudden stop. If you are mounting your changer under the rear deck, make sure it won't interfere with the torsion spring of the trunk lid. Before you choose an exact mounting spot, observe how the spring moves as you close the lid. Don't mount your changer anywhere it would be subjected to direct sunlight or rain.

When drilling holes for mounting brackets, be careful not to drill into the gas tank or a brake line.

Connecting

If you're using a head unit with CD changer controls, you'll need to run the CD changer cable to the head unit and plug it in the back. You'll have to pull the head unit out to do this. If you have a factory head unit with CD changer controls, you'll need to insert an OEM interface CD changer converter between the aftermarket CD changer and the factory head unit.

If you're mounting your CD changer in the trunk, run the CD changer cable down one side of your car, along the door sill. Look for a hole to pass the cable from the trunk into the passenger compartment. You may have to remove the back seat to find it. To conceal the cable under the carpeting, you'll have to take off the door sill plate (or plates).

If you're mounting the changer under a front seat, do not run the cable too close to a seat rail where it could be cut or pinched when the seat is moved.

If you have an amp in the trunk, run the CD changer cable on the side opposite the amp's 12-volt power cable. You'll be less likely to introduce engine noise into your system. Secure any excess cable with wire ties so it doesn't end up in the way of the pedals or hood latch.

After you plug the CD changer cable into the head unit, test the changer thoroughly before you fully reinstall your head unit.

CD Changers with FM Modulators. CD changers with FM modulators require a few additional connections. These can include battery +12 volts and ignition/accessory +12 volts. The best place to find these power sources is at your fuse panel. Use fuse taps at appropriate locations in the panel. You'll also need to connect the black ground wire to the metal chassis of the vehicle with a screw.

Your factory antenna cable will need to plug into the FM modulator (usually integrated into the changer, or a separate module in some units).

A cable from the FM modulator also needs to plug into the head unit's antenna jack. When deciding where to mount a separate FM modulator module, be sure its antenna cable will reach your head unit's antenna input. Use cable ties to secure any slack in the antenna lead, so it can't fall down and interfere with your pedals.

Try to make the selector switch for the FM modulator frequency as accessible as possible. (Some models allow the frequency to be changed from the dash-mounted control panel.) If the selector switch will be difficult to access, set it before you mount the unit. Determine which of the available FM frequencies is least likely to be used by a strong local radio signal, and choose that one.

CHAPTER **10**

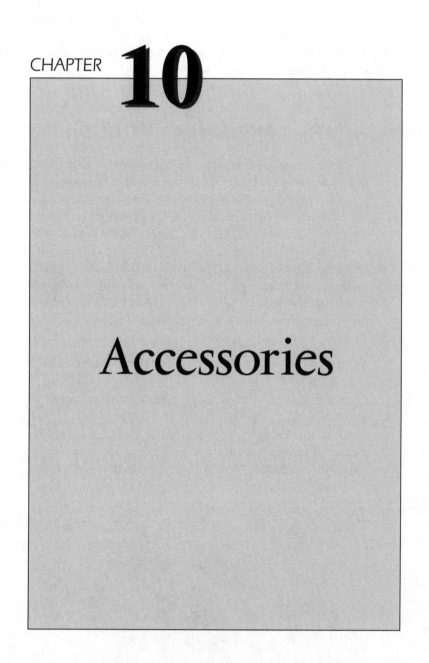

Accessories

The level of hype for accessories is even higher than that for car stereo as a whole. This section will help you separate science from snake oil when it comes to items such as power line caps, premium speaker wire, and sound deadeners.

Power Line Capacitors

The term *power line capacitors* refers to one or more large capacitors (0.25 farads or greater) connected to an amplifier's power wire (Fig. 10-1). The purpose of using these is to provide a sort of reserve power source from which the amplifier can rapidly draw power when it needs it (such as during a deep bass note). The electrical theory is that when the amplifier attempts to draw a large amount of current, not only will the battery be relatively slow to respond, but the voltage at the amplifier will be a little lower than the voltage at the battery itself (this is called *line drop*). A capacitor at the amplifier will try to stabilize the voltage level at the amplifier, dumping current into the amplifier as needed.

Should You Install a Power Line Cap?

Considering the relatively high cost of power line caps, you should first make sure that you have done everything possible to reduce the voltage drop between the battery and the amplifier. This means keeping power and ground cables as short as possible, using the proper wire gauge, and soldering crimped connectors to reduce their resistance.

Figure 10-1
Power line caps.
(*Courtesy of MCM Electronics.*)

If you've done what you can to reduce line drops, and your dash lights still dim every time your subwoofer hits a strong bass note with the music turned up and your engine running, a power line cap could help you out.

Choosing the Right Cap

The commonly accepted rule of thumb for determining proper capacitor size is 1 farad per kilowatt of amplifier power. For example, a system running at 250 watts would need a 0.25-farad (or 250,000-μF) capacitor. Choose a cap with a working voltage (WVDC) rating of 20 volts. Anything less than this is risky, and anything more is a waste of money.

You should consider the Equivalent Series Resistance (ESR) of the capacitor. Capacitors should have an ESR less than that shown in Table 10-1 for maximum effectiveness of the cap.

Equivalent Series Inductance (ESL) is also important, because it dominates the impedance of the capacitor at the high frequencies used by switching power supplies in high-power amps. For a capacitor to provide a benefit at high frequencies, it should have an ESL lower than that of the power cables to the amp. A reasonable value is 1 microhenry. Unfortunately, ESL is not normally advertised.

Another factor to consider is temperature rating. Power line caps have a temperature rating anywhere from 85 to 105°C. No one expects to operate a power line capacitor at temperatures near the boiling point of water, but temperature rating is actually an accurate indicator of the useful lifespan of your power line cap. This is because high temperatures accelerate the process of the electrolyte inside drying out, reducing the capacitance and dramatically increasing the ESR. A cap with a high temperature rating is less susceptible to this process and has a longer lifespan. Opt for a cap with a temperature rating of 95°C or higher.

TABLE 10-1

Maximum Recommended ESR for Power Line Caps

Capacitance	Maximum ESR
0.25 farad	8 mΩ
0.5 farad	4 mΩ
1 farad	2 mΩ
2 farads	1 mΩ

If cost is no object, you can buy electronically monitored power line caps. These provide an indication of the state of charge on an LED bar graph display. Some electronically monitored models also provide reverse polarity protection in case you connect the cap backward.

Installation

If you connect a power line cap backward you will destroy it (the electrolyte will ooze out), so double-check your polarities before making any connections. To install a capacitor, you should not simply attach it to the power and ground wires near your amplifier, as it will draw a very large amount of current from the battery and blow the fuse. Instead, precharge the cap using this procedure:

1. Connect the negative lead to the negative terminal of the cap, but do not connect the positive lead to the positive terminal yet.

2. Insert an automotive lightbulb between the positive lead and the positive terminal of the capacitor to pre-charge the cap. When the lightbulb goes out, the capacitor is done charging.

3. Now connect the positive lead to the positive terminal of the cap.

Battery Savers and Monitors

If you like to crank your stereo for extended periods with the engine off, then you need to be concerned about draining the battery past the point where you can start the car. One deep discharge also sacrifices from one-fourth to one-third the life of your battery.

Some newer cars may keep the battery's voltage from dipping too low, but if your car just sits there as the battery drains, you might consider a battery saver. A battery saver is a computerized switch that automatically disconnects the battery before it's drained to its no-go threshold. Open a door or press a reset button (depending on brand) to reconnect the battery after the battery saver disconnects it. Battery savers are mounted on the side of the battery, and require only a wrench for installation.

A leading consumer test magazine recently compared the Priority-Start! (about $80; Fig. 10-2) and Battery Buddy (about $40) battery savers. Testers simulated a situation in which the driver had left the lights on, then tried to start the car at 0°F. PriorityStart! left the battery with plenty of power to start the engine, but the battery capacity left by the Battery Buddy was sometimes inadequate to start a car.

Another factor important to owners of high-powered car stereos is the contact resistance of a battery saver. This directly translates to voltage drop that can rob your stereo of headroom. The contact resistance of Priority-Start! is less than 1 milliohm, which would cause less than a 25-mV line drop when using a 100-watt per channel amp. The contact resistance of some battery savers may be as high as 50 milliohms, which could cause a line drop of over 1 volt when using a 100-watt per channel amp.

A cheaper alternative is the "Bat Alert" battery monitor (about $15) from Pacific Accessory Corporation (PAC). (See Fig. 10-3.) This device is mounted under the dash and connected to the ignition/accessory 12-volt supply. Its LED is green for battery voltages above 12 volts, yellow between 11.9 and 11.5 volts, and red below that. A red LED tells you it's time to start your engine or hook up a charger. The disadvantage of this approach is that you need to keep your eye on the LED.

Premium Speaker Wire

If you believe the advertising hype, premium speaker wire offers you greater overall depth and clarity, more dynamic range, and tighter, deeper bass.

What the advertising doesn't tell you is that these benefits come from using the proper wire gauge rather than from time-correct windings, magnetic flux tube construction, linear polyethylene dielectric, and so

Figure 10-2
PriorityStart! mounted on battery. (*Courtesy of BLI International.*)

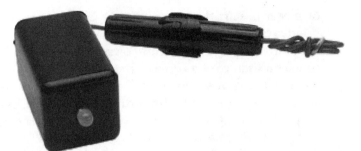

on. If you use the proper wire gauge, then the benefits of premium speaker wire are purely cosmetic.

For speaker wiring, the dominating factor for selecting wire gauge is resistance. For 4-ohm speakers, the speaker wires should have a resistance of less than 0.1 ohms for best performance. Values higher than this can start to affect bass response by allowing the impedance characteristic of the speaker to come into play. A 4-ohm speaker can easily have a 20-ohm impedance peak at its resonant frequency.

Resistance is strictly a matter of wire gauge and length. Table 10-2 provides minimum recommended wire gauges, depending on wire length, for all power amps. The wire lengths shown in Table 10-2 guarantee a resistance of less than 0.1 ohms. Using thicker wire (smaller gauge) and shorter lengths will give you lower resistance.

If you are running 2-ohm loads, subtract 2 from the recommended wire gauge in the chart. For example, use #10-2 up to 30 feet.

Premium Patch Cables (Interconnects)

Interconnects is just a fancy name for patch cables, to make you feel better for paying a premium price. What you get for the extra money is largely cosmetic (Fig. 10-4).

TABLE 10-2

Minimum Recommended Gauge for Speaker Wire

Amplifier	Speaker Wire	
All wattages	up to 12′	#16-2
	up to 20′	#14-2
	up to 30′	#12-2

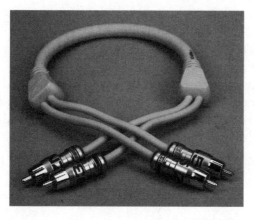

Figure 10-4
Premium patch cables.
(*Courtesy of MCM
Electronics.*)

Factors such as characteristic impedance and low-capacitance dielectric just don't make any difference in audio patch cables. These factors become important at higher frequencies, in applications involving extremely long cables, or where impedance matching is used. None of these apply here. There is no scientific evidence that oxygen-free copper conductors provide any audible benefit whatsoever.

Gold-plated connectors do provide an additional degree of corrosion resistance, but you can buy patch cables with gold-plated ends without spending a fortune.

One area where there can be an important difference between cables is noise immunity, especially in the noisy automotive environment. This is of most concern where:

■ Long cables are used (such as between the head unit and the trunk)

■ A low (less than 2-volt) preamp-level signal source is involved

■ High-power (greater than 100W × 2) amplifiers are used

Unfortunately, cable manufacturers don't directly specify noise immunity. The best you can do is guess, based on terms like *double-shielded* or *100% foil shield*. Experiments by the Autosound 2000 test lab provide evidence that unshielded twisted-pair cables may furnish the best noise immunity in the automotive environment. Note: System noise performance is often determined by factors other than the type of cable used. See Chap. 11 for more details.

A useful feature of some premium cables is an extra embedded wire for remote amp turn-on. This saves you from having to run a separate wire from the head unit to the power amp.

Portable CD Player Head Unit Adapters

A head unit adapter lets you listen to your portable CD player through your car stereo. This is a reasonable solution if you like your head unit but lack CD capabilities. It's a great solution for leased or rental cars. A few head units have a miniplug input, in which case no adapter is required.

There are two types of adapters available: cassette adapters and FM transmitters (Table 10-3).

Cassette Adapters

A cassette adapter looks like a cassette tape that it has a cable coming out with a plug on the end (Fig. 10-5). These adapters are inserted into your tape deck like a normal tape and the plug is inserted into the headphone jack of your portable CD player. Cassette adapters are available for less than $20.

A disadvantage of using cassette adapters is that the tape deck transport mechanism is active and subject to normal wear during use with the adapter. (The tape head may or may not be subject to wear, depending on the design of the adapter.)

Some inexpensive cassette adapters may not work with all tape decks. For example, Delco tape decks will shut down if they sense that there is abnormal tape motion—not all adapters have what it takes to fool the deck. Choose a reputable brand of adapter to be safe.

FM Transmitters

This adapter is basically a miniature radio station (Fig. 10-6). It converts the audio from the CD player's headphone jack to an FM stereo signal picked

Tip: If you plan to use a portable CD player in your car on a regular basis, you will want to use a power converter to save on batteries. These plug into the lighter jack and convert 12 volts to the voltage needed by your CD player.

Important: Power converters not designed to go with your CD player may not be compatible, even if the connectors are the same. Always check for the correct voltage AND correct polarity before plugging a power adapter into your CD player. Failure to do this can result in a destroyed CD player.

TABLE 10-3

Portable CD Player Head Unit Adapters

Adapter	Comments
Cassette adapter	Requires cassette head unit—wired solution.
	Tape deck transport is active during use and subject to normal wear.
FM transmitter	Uses FM radio—wireless solution.
	Very limited transmission range.

Figure 10-5
RCA cassette adapter.
(*Courtesy of MCM
Electronics.*)

up by your head unit. These are available for about $40 from Radio Shack and other sources.

Because of the heavy compression used by commercial broadcasters to reach the maximum audience, the audio quality using an adapter can be much better than a normal FM broadcast.

The biggest problem with FM transmitters is their very limited transmission distance. Because of strict FCC limits on maximum transmitted signal strength, their effective range is quite limited. For some vehicles, this doesn't present a problem. For others, it's virtually impossible to achieve good reception.

Antennas and Antenna Boosters

If your car already has a working antenna, then there's probably no reason to replace it. The two exceptions to this are if you have an embedded windshield antenna or if you want to convert to a motorized antenna.

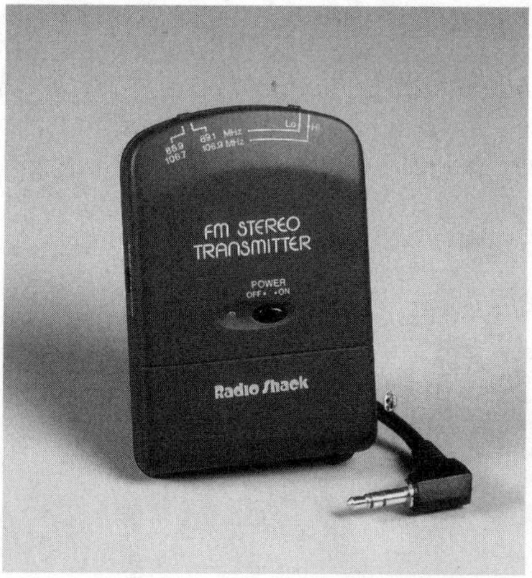

Figure 10-6
Radio Shack FM transmitter (12-2051).
(*Courtesy of Radio
Shack.*)

Embedded Windshield Antennas

Embedded windshield antennas keep your car's body lines clean and they can't get broken off by vandals or at car washes, but they are very directional. This means their performance is good in some directions and poor in others. If you want to pull in weaker stations from all directions, a mast-type antenna is a must.

Motorized Antennas

Motorized antennas (Fig. 10-7) are automatically raised when the radio is on and lowered when it's off. They keep your car's body lines clean, provide good reception in all directions, and only get broken off if you forget to turn off your radio at the car wash. The performance of motorized antennas is comparable to that of fixed mast antennas.

Because of the additional bulk of the motor assembly, it may be difficult or impossible to install a motorized antenna in the front fender of every vehicle. If you mount it in the rear fender, you'll need to use an antenna extension cable and run it under the carpet to the receiver. Some other disadvantages are cost and reliability. The expected life of a motorized antenna is only five to ten years.

Antenna Installation

If you are replacing an existing antenna, attach a length of thin wire to the end of the old antenna cable before you pull it out. As you pull the cable through the car, the wire will snake through the cable's mounting path. Once you've pulled the cable all the way out, attach the wire to the end of the new antenna's cable, then have someone feed the new cable as you pull the wire back into the car.

Figure 10-7
Motorized antenna.
(*Courtesy of MCM Electronics.*)

Antenna Adapters

If you are adding or changing an antenna or replacing a factory head unit, you'll quickly discover that every antenna doesn't directly connect to every head unit. That's where antenna adapters come in. Aftermarket head units and antennas all use the standard Motorola connector, but many factory systems do not (Table 10-4). Adapters are available to adapt aftermarket head units to factory antennas and also to adapt aftermarket antennas to factory head units.

TABLE 10-4

Types of Antenna Connectors

Type	Comments
Standard (Motorola)	Used on all aftermarket equipment.
GM mini	Used on 1988 and up GM.
Ford	Used on some 1995 and up, usually for radios with DSP controls.
Nissan	Diversity (two-antenna) system.
Volkswagen/Euro	Factory antenna needs 12 volts to turn on impedance-matching circuit.

If you're installing an antenna in a vehicle that didn't have one before, scout out the antenna location and cable path the car's designers intended. Look for a rubber or plastic plug on the front fender. Another likely location on Japanese cars is the front pillar between the windshield and front door. Do not run the antenna cable through a hole with other wiring or you will risk picking up interference from those wires.

In the unlikely event that you don't find an existing mounting hole, you'll need to drill one. Check where other cars of the same make have theirs. Once you've chosen a location, protect the area with a piece of duct tape. Use a center punch to mark it and drill a $\frac{1}{8}$-inch pilot hole. Gradually enlarge the hole with successively larger bits to reduce the chance of the bit slipping. Check the hole as you approach the correct size to avoid drilling too large a hole.

Antenna Boosters

Antenna boosters are low-noise amplifiers designed to improve weak signal reception. They are inserted between the antenna and the head unit by plugging the antenna into the booster and the booster output into the head unit.

Antenna boosters can improve *sensitivity,* but not *selectivity.* What this means is that if you are trying to pick up a weak signal in the city (where there are many nearby strong signals), a booster won't help a bit. On the other hand, if you are out in the country trying to pick up a weak signal among other weak signals, a booster might help. How much it might help depends on how good the booster is and how bad your head unit front end is. A cheap booster can actually degrade the weak signal performance of a top-notch receiver.

Sound-Deadening Materials

Sound-deadening materials are used by all automobile manufacturers to reduce the ambient noise level in the passenger compartment. Effective sound deadening means you don't need to turn up the volume control as much to overcome road noise. It also means you'll be able to experience more of the dynamic range available with CDs and even cassettes—you'll be able to hear more of the subtle details in the music. Unlike other accessories, sound deadening provides a benefit even when the stereo is off, by reducing driver fatigue due to noise.

Whether or not a particular vehicle would benefit from additional sound deadening depends on how much soundproofing was done by the vehicle manufacturer. Luxury cars tend to have lots of soundproofing, and are not normally good candidates for improvement. Less expensive cars, older cars, and sports cars are better candidates.

Soundproofing can be an expensive and time-consuming proposition. Materials generally run $1 to $5 per square foot. Dynamat™, Stinger Roadkill™, and Accumat™ (Fig. 10-8) are some of the popular product lines. Each line contains a number of materials targeted for specific applications such as on floor panels or under the hood. Most materials take the form of adhesive-backed sheets, but spray-on products are also available for hard-to-reach areas.

It's important to focus your efforts on the weak link in noise. Otherwise you'll be spending your time plugging pinholes when there's a hole the size of a grapefruit nearby. Determining the weak link can be difficult, but try to identify the source and location of noise as you drive on the highway. Try to judge whether engine noise, noise from the tires on the pavement, or wind noise is dominating. Is most of the noise coming

Figure 10-8
Accumat sound
insulation. (*Courtesy of
Scosche.*)

through the floor, through the doors, or elsewhere? In many ways this is a job for an acoustic engineer with sophisticated measuring equipment and lots of time. In addition, many noise problems, such as wind noise due to the aerodynamics of the car, are difficult or impossible to correct yourself.

The advertising for noise-deadening products generally claims that improvements of 3 dB are possible. This is a noticeable improvement for most people. You need to judge whether the cost of material and your time are worth it.

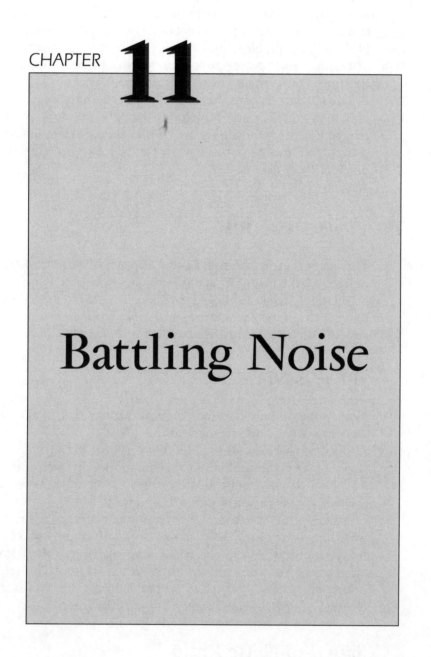

Battling Noise

Most people don't think about noise until after they've installed their system and experience a problem. For basic low-power systems, noise is less likely to be a problem, and this is a reasonable approach. But if you're installing a high-power multiamp system, the best time to think about avoiding noise is when you're planning your system.

This chapter explains the common causes of automotive noise problems, how to avoid them, and how to fix them. Fixing noise problems can be frustrating, time-consuming work. Following the techniques explained in this chapter can help you systematically solve problems and eliminate much of the frustration.

Noise Basics

This section explains some basic principles of noise and the important noise mechanisms in car audio. Reducing noise is a complex subject and many highly technical books have been devoted to it. Understanding the basics will go a long way toward helping you reduce noise when you design, install, and adjust your system or have to troubleshoot it.

What Is Noise?

Noise is the undesired sound you hear during (what should be) silent passages in music or other programming.

Thermal noise sounds like the hiss you hear between stations on your radio. It occurs due to random electron motion in electronic components. It's called thermal noise because the amount of electron motion (and therefore noise) depends on the temperature.

Induced noise can sound like clicking, popping, whirring, buzzing, whistling, or whining. What it sounds like is often a good clue as to its source. Induced noise always has a *source* (such as the alternator), a *channel* (such as a power wire), and a *receiver* (such as an amp). This is important to remember because induced noise problems can be tackled at the source, channel, receiver, or some combination of the three.

Signal-to-Noise Ratio

When dealing with noise, it usually makes sense to think in terms of the signal-to-noise ratio (SNR) rather than the absolute noise level. The signal-to-noise ratio is a measure of the degree of contamination of a signal by noise.

The signal-to-noise ratio is the signal voltage level divided by the noise voltage level. This number is usually expressed in decibels. For example, a 1-volt signal contaminated by 1 millivolt (0.001 volt) of noise would have a signal-to-noise ratio of 1000:1 or 60 dB. To put this in perspective, the signal-to-noise ratio of a Dolby B cassette recording is usually between 60 and 65 dB. Every 10-dB improvement in signal-to-noise ratio (higher is better) sounds like cutting the noise volume in half.

An important property of the signal-to-noise ratio is that it is not changed by amplification or attenuation. For example, suppose a 1-volt signal contaminated by 1 millivolt of noise is fed into an amplifier with a voltage gain of 10. The amplifier output would have 10 volts of signal contaminated by 10 millivolts of noise. The SNR is 1000:1 (or 60 dB) at the input as well as the output of the amp.

Signal Level and Signal-to-Noise Ratio

Even though the signal-to-noise ratio is not directly changed by amplification or attenuation, amplification and attenuation play crucial roles in determining the overall system signal-to-noise ratio. That's because the signal level at each point in a system determines the noise immunity at that point. The higher the signal level, the higher the noise immunity at that point.

When it comes to adjusting your system or choosing equipment, higher signal levels are always better for noise. But you need to make sure that levels aren't set too high, or you'll end up with distortion. The proper procedures are explained later on, in the section on installation and adjustment.

Figure 11-1 shows how high preamp levels improve noise immunity.

Systems (a) and (b) are identical, except that (a) uses a head unit with a 1-volt preamp output level and (b) uses a head unit with a 4-volt level. In both cases, we'll assume that 1 millivolt of noise is induced into the cable between the head unit and amp. We'll also assume that in both cases the amp level controls were properly set to provide full output power with full input voltage from the head unit.

The signal level coming out of each head unit determines the noise immunity at the cables. Even though the induced noise into each cable is the same for both systems, the signal-to-noise ratio of system (b) is 4 times better at the amp input, due to the higher signal level. Since the signal-to-noise ratios are unchanged by the amplifiers, system (b) provides a 4 times (12 dB) better signal-to-noise ratio at the speaker.

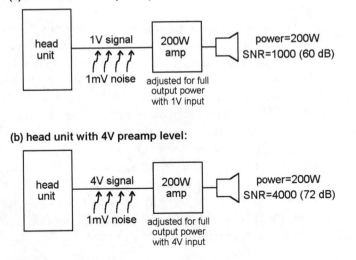

Figure 11-1

How high preamp
levels improve noise
immunity.

Why High-Power Systems Are More Noise Prone

Figure 11-2 shows why high-power systems are more prone to noise. This explains why it's important to think about avoiding noise beforehand when you're planning a high-power multiamp system.

Diagram (*a*) shows a 50-watt system with a head unit having a 1-volt preamp level. Diagram (*b*) shows a 200-watt system with the identical head unit, but a more powerful amp. In both cases, we'll assume that 1 millivolt of noise is induced into the cable between the head unit and amp. We'll also assume that in both cases the amp level controls were properly set to provide full output power with full input voltage from the head unit.

Since the signal level coming out of each head unit is 1 volt and the noise induced into each cable is 1 millivolt, both systems have a signal-to-noise ratio of 1000:1 (or 60 dB). The difference between the two systems is strictly output power. With a 1-volt input, the 50-watt system provides 50 watts to the speaker and the 200-watt system provides 200 watts to the speaker. This makes perfect sense, because we expect to be able to achieve full amplifier power with full input voltage from the head unit.

Now let's see what happens when we adjust the volume control of the 200-watt system to match the output power of the 50-watt system. This is shown in diagram (*c*). (To produce a 50-watt output, a fourfold reduction in power is needed. Because power is proportional to voltage squared, a twofold reduction in voltage at the head unit is required.) Since the signal

Figure 11-2

Why high-power
systems are more
prone to noise.

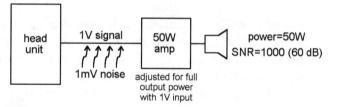

(a) 50 watt amp operating at full volume:

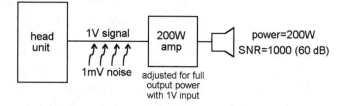

(b) 200 watt amp operating at full volume:

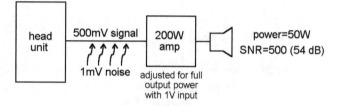

(c) 200 watt amp operating at 50 watt volume level:

coming out of the head unit is now 500 mV, and the noise induced in the cable remains at 1 mV, the signal-to-noise ratio drops to 500:1 or 54 dB. Thus, the signal-to-noise ratio of the high-power system is inferior *when comparing at the same output power.*

Of course, the higher-power system is capable of 4 times more output power, but for normal listening levels you pay a price in noise immunity.

Differential Versus Single-Ended Signals

Using differential (or balanced) signals provides a huge improvement in noise immunity. It's no wonder that all Bose and Ford Premium sound systems rely on differential signaling to provide clean signal transfer.

Unfortunately, most aftermarket equipment is single-ended. It is possible to buy a differential head unit and amp. You can also buy a differential line driver/receiver for use with single-ended equipment. Another often overlooked possibility: The speaker-level outputs of high-power head units are differential. Using the speaker-level inputs of an amp or crossover with true differential inputs gives you a big noise advantage.

Here's how differential signaling works (Fig. 11-3): The differential outputs of a head unit (or differential line driver) provide two equal but opposite versions of the music signal—an inverted version and a noninverted version. At the differential inputs of the amplifier (or differential receiver), the two versions of the signal are subtracted. Since any noise presumably affects both versions of the signal the same, subtraction causes cancellation of the noise. Because one of the music signals was inverted, subtraction also results in a music signal that is twice the original signal.

Theoretically, differential signaling can result in complete cancellation of induced noise. In practice, nonidentical noise into both wires and imbalances in the subtracter limit the effectiveness of the cancellation process. It's very important in differential applications to make sure both wires are receiving identical noise—twisted-pair cabling is often used for this purpose.

Ground Loops

Ground loops are one of the major causes of system noise problems in vehicles. It's important to understand the structure of ground loops and why they are an invitation to noise.

Basically, a ground loop is a loop formed by redundant ground wiring. For example, ground loops are formed by the chassis ground and signal ground wiring between a head unit and amp as shown in Fig. 11-4.

Why is a ground loop a problem? One of the fundamental laws of electromagnetics (Faraday's law) states that when a changing magnetic field passes through a loop of wire, a voltage is induced in the wire. The automotive environment is *loaded* with changing magnetic fields (and they're

Figure 11-3
How differential signaling works.

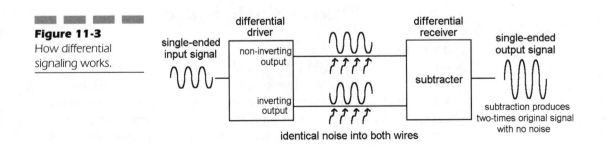

single-ended input signal

differential driver
non-inverting output
inverting output

identical noise into both wires

differential receiver
subtracter

single-ended output signal

subtraction produces two-times original signal with no noise

Figure 11-4

Anatomy of a ground loop.

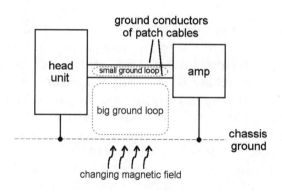

changing magnetic field

all potential noise sources). When a changing magnetic field passes through a ground loop, a noise voltage is induced across the ground conductor.

How do you fight against ground loops? For one thing, avoid them as much as possible. Amps, equalizers, and crossovers with true differential inputs (either preamp level or speaker level) break ground loops because they do not create an input connection to ground. IMPORTANT: This strategy applies to single-ended source applications too—not just differential. You don't need a differential output to take advantage of the ground loop breaking ability of a differential input.

You can also buy true differential line output converters and preamp-level ground loop isolators for installation at the input of an amp having ground-referenced inputs.

For short runs between two pieces of equipment, it usually suffices to make the loop area of ground loops as small as possible. This is because the amount of noise voltage induced in the loop depends on the size of the loop. Make the loop small by minimizing both the length and width of the loop. (Minimizing the width of the loop implies running your cables close to the chassis ground.) The amount of noise voltage induced in the loop also depends on the strength of the changing magnetic field through the loop. This means keeping the loop away from noise sources such as motors and other wiring.

Common Ground Impedance

In the preceding discussion, you may have wondered about simply cutting redundant ground conductors as a cure for ground loops. In the previous example, you could have cut the ground conductors of the patch cables and used chassis ground for signal ground as well. This would have eliminated the ground loop problem, but it would have introduced another noise problem—one due to common ground impedance (Fig. 11-5).

The chassis ground is shared by virtually every accessory and electrical device in the vehicle. This means that return currents from all these potential noise sources flow through the chassis. Even though the chassis resistance is quite low, it's not zero. Return currents from potential noise sources flow through the resistance between two points on the chassis and produce a noise voltage between them. In this example, the noise voltage would directly couple to the inputs of the amplifier.

As Richard Chinn of AudioControl likes to say, "Ground isn't ground!"

Power Supply Noise

Another way that noise can find its way into your stereo system is through the 12-volt power supply.

The alternator is a primary source of noise on the 12-volt line. It supplies raw electrical power for the vehicle and is used to charge the battery. It produces alternating current pulses that are rectified and coarsely regulated. The battery provides strong filtering action for alternator-based noise, but there is still plenty to go around.

Electrical accessories can also introduce large amounts of noise onto the 12-volt line. Accessories actually create noise on the 12-volt line by *using* power. This is analogous to the water system in your home, where flushing a toilet causes a pressure drop in the shower. Devices that switch large amounts of supply current off and on are prime offenders. Good examples are motors, horns, turn signals, and brake lights.

Car stereo equipment is designed to be somewhat immune to noise on the power line, but high levels of power line noise can cause objection-

Figure 11-5
Common ground impedance.

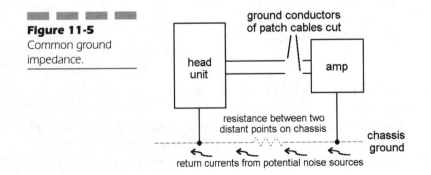

able noise at equipment outputs. All equipment is not created equal with respect to power line noise immunity, but manufacturers generally don't give you any specs to compare in this area.

Power supply noise problems are usually dealt with by adding an appropriate power line filter at either the source of the noise problem or the equipment that is affected by it.

Buying Equipment

When it comes to buying equipment with an eye toward good noise immunity, the most important factors are high preamp levels and ground loop prevention. Choose amps, crossovers, and equalizers with differential inputs (speaker level or preamp level) for ground loop isolation. For competition-caliber systems, there are a number of specialty products to keep even the most complex and powerful systems quiet.

Head Units

If you plan to use the preamp-level outputs of your head unit, the most important thing you can do to improve noise immunity is to choose a model with high preamp output levels. (Alternately, you can use the speaker-level outputs of the head unit for driving an amp, crossover, or equalizer. The high voltage levels and differential drive provided by high-power speaker-level outputs is actually superior to 4-volt preamp outputs for noise immunity.)

Unfortunately, most aftermarket head units have single-ended preamp-level outputs. An exception is the Kenwood eXcelon KDC-PS909 (Fig. 11-6). You can select unbalanced (4 volts) or balanced (8 volts) preamp output signals for a better signal-to-noise ratio. (Using the balanced signal requires an amp that accepts balanced inputs.)

Figure 11-6
Kenwood eXcelon
KDC-PS909. (*Courtesy of Kenwood.*)

Amplifiers

When it comes to buying an amp, there are two factors to consider from the standpoint of system noise. The first is whether the amp has differential inputs and the second is the amp's signal-to-noise ratio.

Differential inputs are the key to avoiding ground loops with both differential and single-ended sources. Both preamp-level and speaker-level inputs can be single-ended or differential, so make sure the inputs you plan to use are differential. Differential inputs should be considered a must for high-power amps located more than a foot or two from a preamp-level component feeding them. More and more amplifiers now feature differential inputs to reduce noise—there's no good reason not to choose a model with this feature.

An amp's signal-to-noise ratio spec is of lesser importance. Almost all decent amplifiers have about 100 dB signal-to-noise ratio, which is more than enough for most applications. In most systems, the signal-to-noise ratio is dominated by induced noise prior to the amp or the SNR of the source itself.

Crossovers and Equalizers

As with amps, the two factors to consider from the standpoint of system noise are differential inputs and signal-to-noise ratio.

Differential inputs are the key to avoiding ground loops with both differential and single-ended sources. Both preamp-level and speaker-level inputs can be single-ended or differential, so make sure the inputs you plan to use are differential. If the crossover or equalizer you like doesn't have differential inputs, you may need to resort to a ground loop isolator later on.

The signal-to-noise ratio spec is of lesser importance. Almost all decent crossovers and equalizers have signal-to-noise ratios of about 100 dB, which is more than enough for most applications. In most systems, the signal-to-noise ratio is dominated by induced noise prior to the crossover or the SNR of the source itself.

Cables

The type of cable you use can affect the level of induced noise. Cable noise immunity is most important where:

- Long cables are used (such as between the head unit and the trunk).
- A low (less than 2 volts) preamp-level signal source is involved.
- High-power (greater than 100W × 2) amplifiers are used.

Unfortunately, cable manufacturers don't directly specify noise immunity. The best you can do is guess, based on terms like *double-shielded* or *100% foil shield*. Experiments by the Autosound 2000 test lab provide evidence that unshielded twisted-pair cables may furnish the best noise immunity in the automotive environment.

Line Drivers

A line driver boosts the signal level and often provides a balanced output, both of which greatly improve noise immunity of a preamp-level signal over a long cable. AudioControl's Overdrive™ line driver (Fig. 11-7) does both of these and more.

The Overdrive provides differential inputs for ground loop isolation, adjustable gain up to 24 dB, and a peak signal voltage of 13.5 volts. It is capable of driving the high capacitive loads presented by some cables.

Master Volume Control

For competition-grade systems involving lots of processors and high-power amplifiers, you may have to resort to some heavy artillery in the battle against noise. The AudioControl MVC™ (Master Volume Control) is a prime example (Fig. 11-8).

The MVC is installed in the system's preamp-level signal path as the very last component before the amps. (In fact, the physical distance

Figure 11-7
AudioControl
Overdrive line driver.
(*Courtesy of*
AudioControl.)

Figure 11-8
AudioControl MVC.
(*Courtesy of*
AudioControl.)

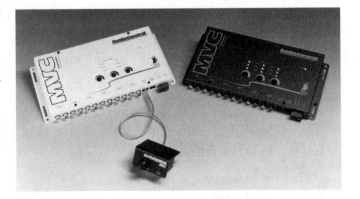

between the MVC and the amps should be as small as possible.) It controls the system volume by attenuating the signal level of each channel just before it enters the amp. This allows the head unit volume control and the processor gain settings to be optimized for maximum signal-to-noise ratio at all times, allowing the audio signal to be sent through all the processors and signal wires at the highest undistorted level possible. A single rotary knob mounted at the dash controls the system volume.

The MVC is a six-channel (three stereo pair) device. Additional MVCs can be connected together should you need more than six channels. The MVC includes a remote out connection that can be used to delay the turn-on of the amps by 1.5 or 4 seconds (selectable). This allows the head unit and signal processors to completely power up before the amps turn on, eliminating turn-on thumps.

Installation and Level Adjustment

Proper installation and level adjustment are just as important as choosing the right equipment in achieving good noise immunity. Proper installation minimizes the induction of noise into your system and proper level adjustment minimizes its audibility.

Equipment Placement

When deciding where to put each of the components in your system, noise immunity considerations aren't likely to be near the top of your list. But try to keep noise pickup in mind, especially where you have flexibility in mounting locations.

The Science Behind the Master Volume Control

The AudioControl MVC is a perfect example of the maxim: "For best signal-to-noise ratio, maximize the signal level at each point in the signal path." Figure 11-9 shows how the MVC improves the signal-to-noise ratio.

Figure 11-9
How the MVC improves the signal-to-noise ratio.

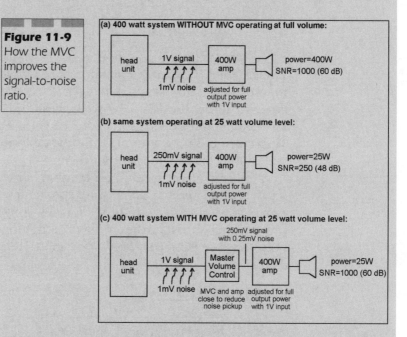

Diagram (*a*) shows a head unit having a 1-volt preamp output level driving a 400-watt amp. There could be any number of processors between the head unit and amplifier input, but we'll simply assume that the noise induced after the head unit reaches a level of 1 millivolt at the amplifier input. At full volume, this system has a signal-to-noise ratio of 1000:1 (60 dB).

Diagram (*b*) shows the same system as diagram (*a*), but with the volume reduced at the head unit. At the amplifier input, the signal level is reduced to 250 mV, but the noise level is unchanged from diagram (*a*) since the volume control setting doesn't affect the noise. This produces a signal-to-noise ratio of 250:1 (48 dB). The situation only gets worse as the head unit volume control is further reduced.

Diagram (*c*) shows the same system, but with a Master Volume Control installed. In this case, the head unit volume control is fixed to maintain a 1-volt level. Now, the MVC is used to change the system volume. Because the MVC attenuates the signal *after* most of the noise induction has occurred, it attenuates the noise as well as the signal. This maintains the maximum system signal-to-noise ratio over the most of the volume control range.

From the standpoint of noise pickup, cable length and cable routing are prime factors. *Equipment placement is important because it dictates cable length and routing.* Choose equipment locations that:

- Allow you to route cables away from noise sources (such as motors and wiring)
- Let you use short cables for highly noise-sensitive signal paths

An example of a highly noise-sensitive signal path would be the low-voltage (500 mV) preamp-level outputs of a head unit. Suppose you want to use these outputs for driving a crossover and amps. Putting the crossover under the dash, close to the head unit, lets you use a short cable for the highly noise-sensitive path. The voltage gain of the crossover provides enhanced noise immunity for the cables to the amps, so cable length and routing is less important for them.

Another example would be a trunk-mounted subwoofer system with separate crossover and amp. Using speaker-level signals to the crossover provides good noise immunity for the long run to the trunk. The preamp-level output of the crossover to the amp is a comparatively highly noise-sensitive path, and should be a short cable.

Cable Routing

Avoid running preamp-level signal cables side by side with high-current power wiring. If you plan to install amps in the trunk, run the signal cables down one side of the car and the power wires down the other.

Try to route signal cables away from electromagnetic noise sources such as motors and other wiring. Where possible, run your cables close to the chassis.

Experiment with different cable routing paths to minimize noise pickup. Start the engine and turn on the head lights and stereo. To make it easier to hear noise, set the sensitivity controls of the amps and processors to max, but turn the volume control of the head unit to minimum. Sometimes moving a cable just a few inches can make a big difference. The best time to experiment with cable routing is during the installation, not after it.

Grounding

The chassis is the generally preferred grounding point for amps and other components because of its low resistance. But it's important to use the proper grounding techniques (Fig. 11-10) to reduce noise pickup due to common ground impedance with potential noise sources.

Figure 11-10(*a*) shows two components sharing a ground wire to the chassis. This allows the ground currents of one component to affect the ground reference voltage of the other component. This is mainly a concern where an amp is affecting the ground reference of a component such as a crossover. At a minimum, each amplifier should have its own connection to the chassis.

Figure 11-10(*b*) shows two components using individual but widely separated chassis grounding points. This allows return currents that flow through the chassis to produce a noise voltage between the two points.

Figure 11-10(*c*) shows the proper grounding technique—each component has its own connection to the chassis, and all connections are closely located. It's okay for low-current components to share a common ground point.

Not all metal in a car is electrically connected to the chassis, so use a multimeter to confirm a low-resistance connection to a known good ground. Look for values less than 0.3 ohms. Note: When making ultra-low resistance measurements, first short the two test probes together and observe the reading. If it's an analog meter, zero the meter at this point.

Figure 11-10
Improper and
proper grounding
techniques.

wrong

**(a) two components sharing a
ground wire to the chassis:**

resistance of
— shared ground wire

chassis
ground

wrong

**(b) ground impedance between
two distant points on chassis:**

resistance between two
distant points on chassis

chassis
ground

right

(c) proper grounding technique:

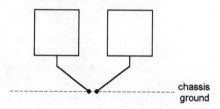

chassis
ground

For a digital meter, you will need to mentally subtract this reading from your measurement readings.

Step-by-Step Level Adjustment

When it comes to adjusting your system, higher signal levels are always better for reducing noise. But you need to make sure that levels aren't set too high, or you'll end up with distortion. The goal is to find the level that maximizes the signal-to-noise ratio while keeping you out of distortion.

Important: Some cars (Audi, Porsche) have galvanized bodies. In these cars, you must find one of the manufacturers' grounding points or else some noise can result.

The procedure given next covers the general case of the cascade of many components, each with a level control at its input (Fig. 11-11). For simplicity, this procedure is written assuming two processors between the head unit and amp. For example, processor 1 might be an equalizer and processor 2 might be a crossover. Your system may have more or fewer components cascaded together, but the process remains the same. Note: Some systems require a deviation from this procedure to satisfy other constraints (such as proper level matching between channels in an amplifier-saving configuration). See the appropriate project chapter when in doubt.

1. Set the amp level control and all processor level controls to minimum.

2. With music, turn up the head unit volume control. When you reach the point where you start to hear distortion, turn the volume control

Important: When mounting a power amp (or any car stereo component), secure it in such a way that the metal case does not make electrical contact with the car chassis. This guarantees that the amplifier's ground wire is the exclusive connection between the amp's internal ground and the system ground. Electrical isolation is important to prevent ground loops.

NOTE A few components have electrically isolated cases—these should be electrically connected to the car chassis to provide shielding to the internal circuitry. When in doubt, use an ohmmeter to measure the resistance between the case and ground wire of the component.

Use rubber grommets inside metal mounting bracket holes to prevent screws from making an electrical connection with the chassis. Alternately, you can mount your components on a board, then mount the board to the car chassis.

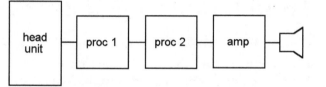

Figure 11-11
The cascade of many
components.

down slightly until the distortion disappears. Leave the volume control at this setting.

NOTE *If the sound from the speaker is so low that you cannot clearly hear the
onset of distortion, turn up the amp's level control as needed. If turning up the
amp's level control to max still isn't loud enough, then start to turn up processor
2's level control as well. Return level control settings on the amp and processor 2
to minimum when you finish this step.*

3. Now, turn up the processor 1 level control. When you reach the point
where you start to hear distortion, turn the processor 1 level control
down slightly until the distortion disappears. Leave the processor 1
level control at this setting.

NOTE *If the sound from the speaker is so low that you cannot clearly hear the
onset of distortion, turn up the amp's level control as needed. Return the amp's
level control setting to minimum when you finish this step.*

4. Now, turn up the processor 2 level control. When you reach the point
where you start to hear distortion, turn the processor 2 level control
down slightly until the distortion disappears. Leave the processor 2
level control at this setting.

NOTE *If the sound from the speaker is so low that you cannot clearly hear the
onset of distortion, turn up the amp's level control as needed.*

5. Turn up the amp level control until the system is as loud as you'll
ever play it or you begin to hear distortion. If you hear distortion,
slightly decrease the amp level control.

Troubleshooting Noise Problems

If you've followed the advice in the previous sections on buying, installing, and setting equipment, then chances are good you won't need this section. But sometimes even well-designed systems have noise problems. The vehicle itself may aggravating the problem—deteriorated ignition or charging system components, a loose battery cable clamp, or missing or defective factory noise suppressors can increase the conducted and radiated noise levels.

This section introduces the most important supplies, tools, and techniques for troubleshooting and repairing noise problems. It concludes with a step-by-step troubleshooting procedure. Fixing noise problems can be frustrating, time-consuming work. Following the techniques explained in this section can help you systematically solve problems and eliminate much of the frustration.

> **Important:** Troubleshooting techniques should be used only *after* you're satisfied that the system level adjustments are right.

Identifying Noise Sources

Knowing what each of the common noise sources sounds like can make the job of troubleshooting easier. This is especially true for AM/FM noise entering the head unit through the antenna or power line. Table 11-1 lists the most common noise sources, what they normally sound like, and what to try to fix them.

Shorting Plugs

Shorting plugs (or muting plugs; Fig. 11-12) are probably the cheapest and most useful troubleshooting aid for system noise problems. Shorting plugs are used for muting the inputs of an amplifier or other component to determine if the noise is entering after that point.

You'll want at least a pair of them and can make them yourself. Make each one by soldering a short piece of wire between the center and outer conductors in a standard RCA plug.

See the step-by-step troubleshooting procedure at the end of this section for full details on using shorting plugs.

TABLE 11-1

Common Noise
Sources

Noise Source	What It Sounds Like	What to Try
Alternator	A whine whose pitch changes with engine speed	Capacitor to ground at the alternator output
Ignition	Ticking that changes with engine speed	Using a sniffer to further localize the source of the problem
Turn signals	Popping synchronized with the turn signals	Capacitor across the turn signal flasher
Brake lights	Popping synchronized with the brake lights	Capacitor across the brake light switch contacts
Blower motor	Ticking synchronized with the blower motor	Capacitor to ground at the motor hot lead
Dash lamp dimmer	A buzzy whine whose pitch changes with the dimmer setting	Capacitor to ground at the dimmer hot lead
Horn switch	Popping synchronized with horn	Capacitor between the hot lead and horn lead at the horn relay
Horn	Buzzing synchronized with the horn	Capacitor to ground at each horn hot lead
Amplifier power supply	A nasty buzz, not affected by engine speed	Grounding the amp chassis

Ground Loop Isolators

Ground loop isolators (Fig. 11-13) are as much a diagnostic tool as they are a repair means. By inserting one in the signal path of a known or suspected ground loop, you can quickly determine if a ground loop involving that path is degrading system noise immunity.

Figure 11-12
Shorting plugs.

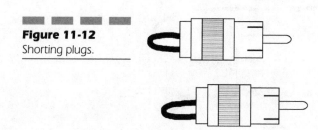

■■ ■■ ■■ ■■

Figure 11-13
Ground loop isolator.
(*Courtesy of Navone Engineering.*)

For diagnostic purposes, any ground loop isolator will do. If you decide you need to install one permanently to solve a system noise problem, then it pays to buy a high-quality model. Low-quality units usually have premature bass response roll-off. Passive ground loop isolators use transformers, and are susceptible to magnetic fields, so you need to keep them away from motors and other wiring. Both of these problems can be avoided by using an active ground loop isolator.

A good way to do an A/B test for the level of noise improvement provided by a ground loop isolator is to use a ground bypass wire (Fig. 11-14).

After you've installed the ground loop isolator, start the engine and turn on the headlights and the stereo. Turn the head unit volume control to minimum. Now use the ground bypass wire to momentarily short the ground connection across the ground loop isolator. You only need to short one channel, not both. If you hear a noise difference, then a ground loop was degrading system noise immunity.

Power Line Filters

Power line filters are used to reduce the amount of noise present on the 12-volt power line. Filters may be used at the source of the noise problem, at the equipment that is most affected by it, or on the power line between the two.

■■ ■■ ■■ ■■

Figure 11-14
Ground bypass wire for ground loop isolator.

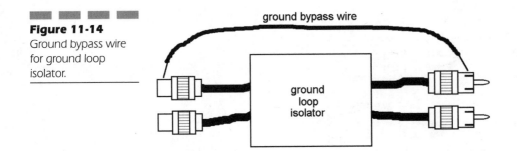

ground bypass wire

ground loop isolator

TABLE 11-2

Types of Power Line
Filters

Type	Connection Diagram	Comments
Capacitor	12-volt line	Can be added anywhere on the 12-volt line with little risk of degrading the noise performance of the system. Solves only a handful of problems.
Choke	12-volt line in · 12-volt line out	Usually provides better noise rejection than a capacitor. Can pick up magnetic fields, especially unshielded models.
Three-terminal passive	12-volt line in · 12-volt line out	Provides filtering action superior to that of a capacitor or inductor alone.
Electronic suppressor ⟨BEST⟩	12-volt line in · 12-volt line out	Provides best noise rejection of all power line filters.

There are four common types of power line filters (Table 11-2): capacitors, chokes (inductors), three-terminal passive filters, and electronic noise suppressors.

Capacitors.   Capacitors (Fig. 11-15) are easily connected from any 12-volt point to chassis ground. The idea behind them is to provide a low-

Figure 11-15
Capacitor. (*Courtesy of Radio Shack.*)

impedance path to ground for the noise. Capacitors used for noise suppression are generally 0.5 µF in value. For good high-frequency performance, electrolytics are not used here.

Capacitors specifically designed for automotive noise suppression often have only a single wire coming from them—the ground connection is made via the mounting tab of the cap. Feed-through designs are also common. These also make the ground connection via the mounting tab, but an in wire and out wire (connected internally) are provided for wiring convenience.

Unlike the other filter types, capacitors can be added anywhere on the 12-volt line with no risk of exceeding the current rating of the filter and little risk of degrading the noise performance of the system. Unfortunately, only a handful of problems are solvable with capacitors—these tend to be noises affecting radio reception.

Chokes. Chokes (inductors; Fig. 11-16) are installed in series with the 12-volt line, normally right at the power lead of the stereo component sensitive to noise. The idea behind them is to provide a high-impedance path for the noise getting to the component. Chokes usually provide better noise rejection than capacitors.

Because all the current used by the stereo component flows through the choke, you need to buy a choke that has a current rating suitable for your application. Current rating is a good indicator of the DC resistance of a particular choke—it's not an indicator of its noise filtering ability. Don't buy a choke with a significantly higher current rating than you need—it can have lower inductance and provide inferior filtering. Since manufacturers don't normally provide inductance specifications for their chokes, beware of unusually small models.

Mount chokes away from motors and other wiring. The unshielded models especially can easily pick up magnetic fields.

Figure 11-16
Choke. (*Courtesy of MCM Electronics.*)

Three-Terminal Passive Filters. Three-terminal passive filters (Fig. 11-17) are installed in series with the 12-volt line, normally right at the power lead of the stereo component sensitive to noise. They require a ground connection as well. Three-terminal passive filters contain both capacitors and inductors. This can provide filtering action superior to that of capacitors or inductors alone. As with chokes, you need to buy a model that has a current rating suitable for your application.

Electronic Noise Suppressors. Electronic noise suppressors (Fig. 11-18) are three-terminal devices installed the same way as three-terminal passive filters. Electronic noise suppressors can provide better performance than even three-terminal passive filters. As with chokes and three-terminal passive filters, you need to buy a model that has a current rating suitable for your application.

Noise Sniffers

Noise sniffers are hand-held electronic devices used to track down sources of radiated noise. These are frequently used by professionals to solve difficult cases.

Figure 11-17
Three-terminal passive filter. (*Courtesy of Navone Engineering.*)

Figure 11-18
Electronic suppressor. (*Courtesy of Navone Engineering.*)

There are two main types of sniffers used in automotive noise hunting. AM/FM sniffers (Fig. 11-19) are used to find sources of radio interference. These are modified radios in which the antenna has been replaced with a probe. Electromagnetic radiation (EMR) sniffers (Fig. 11-20) are used to find sources of audio-frequency magnetic fields. These are modified tape players in which the tape head has been removed and extended with a cable to form a probe. Headphones are preferred with either type of sniffer because they block some of the mechanical sounds of the engine.

The AM/FM sniffer is used by tuning the sniffer to the same frequency at which the problem is noticed on the vehicle's radio. Start the vehicle's engine and probe around the engine compartment until the same noise can be heard through the sniffer's headphones. Move the probe to find the loudest source of the noise.

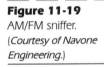

Figure 11-19
AM/FM sniffer.
(*Courtesy of Navone Engineering.*)

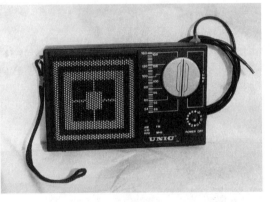

Figure 11-20
EMR sniffer. (*Courtesy of Navone Engineering.*)

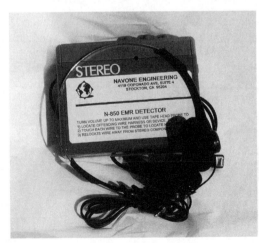

The EMR sniffer is used in a similar fashion. Start the vehicle's engine. Start probing near the cable or stereo component that is picking up the noise. Move the probe to find the loudest source of the noise. The orientation of the probe can make a big difference, so try various orientations. You can use the probe to find the exact wire bundle and often the specific wire causing the problem. Once found, you can shield the wire with copper tape (be sure to ground the tape), or, as a last resort, reroute the offending wire by splicing in an extension.

Copper Shielding Tape

Copper shielding tape (Fig. 11-21) is useful for wrapping around wires and cables to provide shielding. It can be wrapped around patch cables you want to shield from noise, or around power and accessory wiring to reduce radiated emissions. Shielding power and accessory wiring to reduce radiated noise is the generally preferred strategy because there is little risk of making things worse. Shielding patch cables can actually make things worse by coupling chassis noise into them.

Copper shielding tape uses an electrically conductive adhesive to provide a low-resistance path between the backing and substrate. Tinned copper shielding tape is also available to make the job of soldering to it easier.

Any bundles of wire nearby and parallel to patch cabling are good candidates for wrapping with copper tape. When wrapping a bundle of wires

Figure 11-21
Copper shielding tape.
(*Courtesy of MCM Electronics.*)

with tape, there should be a minimum ¼-inch overlap between layers. To maximize the shielding effectiveness, the shield wrap should be electrically grounded to the chassis. Solder short wires to the wrap every few feet and electrically connect them to the chassis via ring crimp connectors and self-tapping screws.

Step-by-Step Troubleshooting Procedure

The key to effective troubleshooting for noise problems is to use a methodical approach to identify the problem. This saves time, frustration, and money.

To troubleshoot a noisy system, follow these steps:

1. Start at an amp where you hear noise. Replace each of the signal cables feeding into the amp with a shorting plug. Start the car's engine, turn on the headlights, and turn on the stereo. Is the result clean? If you hear noise at this point (which is rare), noise is either getting in on the amp's power supply wire, there is something abnormal about the speaker load (e.g., a speaker wire is shorting to the chassis), or the amp is defective.

2. If the result is clean, then it's time to move the shorting plugs upstream (toward the head unit). Turn off the stereo first. For this example, let's assume a crossover is located between the head unit and amp. Plug the cables from the crossover back into the amp. Unplug the cables from other outputs of the crossover for the moment. Now replace each of the signal cables feeding into the crossover with a shorting plug. Start the car's engine, turn on the headlights, and turn on the stereo. Is the result clean? If you hear noise at this point, and your amp lacks differential inputs, the noise is probably due to a ground loop involving the cables between the crossover and amp. (This theory can be tested by temporarily inserting a ground loop isolator at the amp's inputs.) It could also be due to a high level of induced noise on the cable or noise getting in on the crossover's power supply wire.

3. If the result is clean, replace the cables you just unplugged from the other outputs of the crossover. If noise returns, you've got a ground loop involving the cables you just plugged in. If it's still clean, then the problem has to involve the head unit connection to the crossover. If your crossover lacks differential inputs, the noise is probably due to a ground loop involving the cables between the head unit and

crossover. (This theory can be tested by temporarily inserting a ground loop isolator at the crossover's inputs.) It could also be due to a high level of induced noise on the cable or noise getting in on the head unit's power supply wire.

Once you've located the component or cable most affected by the noise and made a preliminary diagnosis of the problem as ground loop related or not, you'll need to follow up each of your leads. If you discovered a ground loop problem, you'll need to break the loop with a ground loop isolator or equipment having differential inputs. If you don't have a ground loop but are experiencing high induced noise in a cable, you'll need to experiment with cable routing and shielding nearby wiring with copper tape. If this fails to fix the problem, the next step is to track down the source of the noise itself. You may have deteriorated electrical system components or inadequate factory noise suppressors. This level of troubleshooting usually requires professional help. Finally, if you suspect your problems are power line related, try an appropriate filter.

RECOMMENDED MAIL-ORDER SUPPLIERS

Crutchfield
1 Crutchfield Park
Charlottesville, VA 22906-9032
(800) 955-3000
www.crutchfield.com

Great selection of quality car stereo equipment combined with unequaled service. Well-trained staff and Master-Sheet instructions provide all the guidance you'll need.

MCM Electronics
650 Congress Park Drive
Dayton, OH 45459-9955
(800) 543-4330
www.mcmelectronics.com

Car stereo supplies and tools, component speakers, boxes, and electronic components. Their huge catalog contains lots of hard-to-find items.

Parts Express
725 Pleasant Valley Drive
Springboro, OH 45066-1158
(800) 338-0531
www.parts-express.com

Car stereo supplies and tools, component speakers, boxes, and electronic components.

Madisound Speaker Components
8608 University Green
P.O. Box 44283
Madison, WI 53744-4283
(608) 831-3433
www.itis.com/madisound

Component speakers and speaker building supplies. A favorite of home speaker building enthusiasts. Top-notch products at wholesale prices.

MFR Engineering
10308 Indian Lake Blvd. South
Indianapolis, IN 46236-8329
www.a1.com/mfr-eng

Subwoofer Design Toolbox™ direct sales.

RECOMMENDED READING

Books and Technical Papers

The Loudspeaker Design Cookbook
Vance Dickason
Published by Audio Amateur
Press
$34.95

If you're serious about designing and building your own subwoofers, this book is unsurpassed. Now in its fifth edition, this book is an industry standard. Includes a new chapter on auto sound.

Designing, Building and Testing Your Own Speaker System
David Weems
Published by McGraw-Hill
$19.95

Another excellent book for subwoofer builders. Now in its fourth edition, this is a completely revised version of Weems's best seller. A hands-on approach to design and construction of speaker systems.

Car Stereo Installation Made Easy
Crutchfield
FREE

Covers installation techniques for basic and advanced car stereo systems.

AudioControl Technical Papers
AudioControl

Cover a range of topics from car acoustics to crossovers. Visit their web site at www.audiocontrol.com for a list of papers or to download PDF versions of many of them.

Magazines

CarSound
6 Manhasset Avenue
Port Washington, NY 11050
(800) 872-0774
www.carsound.com
$11.97/year (6 issues)

Covers car stereo, plus security and navigation. Includes product reviews, buying tips, installation techniques, and new product guides.

Car Audio and Electronics
P.O. Box 58262

Published for the car audio and mobile electronics enthusiast. Includes

Boulder, CO 80322
(800) 289-0611
www.caraudiomag.com
$21.95/year (12 issues)

test reports and custom installations. Annual product directory issue each April.

Car Stereo Review
P.O. Box 57307
Boulder, CO 80322
(800) 365-0809
$24.94/year (10 issues)

From the editors of *Stereo Review.*

Auto Sound and Security
P.O. Box 56952
Boulder, CO 80322
(800) 759-1521
www.mcmullenargus.com/
pub/spoutdr/autosnd
$19.95/year (12 issues)

Speaker Builder
P.O. Box 494
Peterborough, NH 03458-0494
(603) 924-9464
www.audioxpress.com/
magsdirx/spkrbldr
$32/year (8 issues)

Somewhere between a magazine and a technical journal. For those interested in building loudspeakers as well as those who want to understand the technology behind them.

Bibliography

Chapter 3: Speakers and Speaker Projects

Crutchfield. *Car Stereo Installation Made Easy.* 1995.

Crutchfield. *The Complete Car Stereo and Home Audio/Video Catalog.* Winter/Spring 1998.

Kinsler, Lawrence E., et al. *Fundamentals of Acoustics.* 3rd ed. New York: John Wiley & Sons, 1982.

Chapter 4: Subwoofers and Subwoofer Projects

Clark, Richard. "Blocking That Bass." *CarSound,* December 1997, 10–12.

Dickason, Vance. *The Loudspeaker Design Cookbook.* 5th ed. Peterborough, NH: Old Colony Sound, 1995.

Kinsler, Lawrence E., et al. *Fundamentals of Acoustics.* 3rd ed. New York: John Wiley & Sons, 1982.

Putman, Rob. "Beating Bose." *AUTOMEDIA,* September 1997, 36–43.

Rumreich, Mark. "Box Design and Woofer Selection: A New Approach." *Speaker Builder,* Issue 1, 1992.

Chapter 5: Head Unit Projects

Crutchfield. *Car Stereo Installation Made Easy.* 1995.

Crutchfield. *The Complete Car Stereo and Home Audio/Video Catalog.* Winter/Spring 1998.

Crutchfield. *How to Upgrade Your Factory Stereo: OEM Integration Made Easy.*

Putman, Rob. "Beating Bose." *AUTOMEDIA,* September 1997, 36–43.

Chapter 6: Amplifiers and Amplifier Projects

Crutchfield. *Car Stereo Installation Made Easy.* 1995.

Crutchfield. *The Complete Car Stereo and Home Audio/Video Catalog.* Winter/Spring 1998.

Kinsler, Lawrence E., et al. *Fundamentals of Acoustics.* 3rd ed. New York: John Wiley & Sons, 1982.

Putman, Rob. "Beating Bose." *AUTOMEDIA,* September 1997, 36–43.

Chapter 7: Equalizers and Equalizer Projects

Geddes, Earl and Henry Blind. "The Localized Sound Power Method." *J. Audio Eng. Soc.* 34, no. 3, 1986.

Janis, Robert. "Signal Processors: Trends and Merchandising." *Mobile Electronics Specialist,* June 1996, 10–15.

Kinsler, Lawrence E., et al. *Fundamentals of Acoustics.* 3rd ed. New York: John Wiley & Sons, 1982.

Chapter 8: Biamping and Crossovers

Chinn, Richard. *Crossovers and Biamplification,* AudioControl technical paper number 104, 1987.

Chapter 9: CD Changer Projects

Crutchfield. *Car Stereo Installation Made Easy.* 1995.

Crutchfield. *The Complete Car Stereo and Home Audio/Video Catalog.* Winter/Spring 1998.

Chapter 10: Accessories

"Time for a Battery Check?" *Consumer Reports,* October 1997, 23–27.

Crutchfield. *Car Stereo Installation Made Easy.* 1995.

Navone, David and Richard Clark. "Misconceptions in Cabling." *AUTOMEDIA* 2, no. 6, 1997, 29–42.

Chapter 11: Battling Noise

Chinn, Richard. *Level Matching for Autosound...Or Why Is My System So Noisy?* AudioControl technical paper number 103, 1986.

Navone, David. "A Nose for Noise: How to Use Car Audio Sniffers to Detect System Noise." *AUTOMEDIA,* September 1997, 60.

Navone, David. "Troubleshooting: Sound Advice." *CarSound,* December 1997, 86–88.

Navone, David and Richard Clark. "Misconceptions in Cabling." *AUTOMEDIA* 2, no. 6, 1997, 29–42.

INDEX

A

accessories:
  antenna boosters, 247—248
  battery savers and monitors, 240—241
  CD player cassette adapters, 244
  CD player FM transmitters, 244—245
  motorized antennas, 246—247
  power line capacitors, 238—240
  premium patch cables, 242—243
  premium speaker wire, 241—242
  sound-deadening materials, 248—249
Accumat, 248—249
adapters:
  antenna, 150, 247
  cassette, 244
  common ground (*see* OEM integration adapters)
  factory steering wheel controls, 151
  floating ground (*see* OEM integration adapters)
  (*See also* converters)
Add-A-Circuit fuse holder, 15—16
AGC fuses and fuse holders, 166—168
alternator whine, 258—259
amplifiers:
  automatic turn-on wiring, 168—169
  boosting head units with four preamp outputs and four speaker outputs, 171—175
  boosting head units with no preamp outputs and four speaker outputs, 178—181
  boosting head units with no preamp outputs and two speaker outputs, 181—183
  boosting head units with two preamp outputs and four speaker outputs, 175—178
  boosting premium factory sound system head units, 184—189
  bridging, 67—68, 159—161

amplifiers (*Cont.*):
  built-in crossovers, 61
  connectors for power wiring, 170—171
  distribution blocks, 165—166
  fuses, 166—168
  ground connection, 169—170
  ground wiring, 162—171
  maximum power, 156—157
  mounting location, 161—162
  noise immunity considerations, 260
  oscillation with dvc subs, 82
  peak power, 156—157
  power per dollar, 158—161
  power rating methods, 156—157
  power required, 158
  power required for subs, 65—66
  power requirements for biamping, 210—211
  power wiring, 162—171
  premium factory sound systems, 184—189
  RMS power, 156—157
  setting gains, 189—190
  speaker wiring, 171
  two-ohm stable, 159
  wire gauge, 162—164
ANL fuses and fuse holders, 166—168
antenna boosters, 247—248
antennas:
  adapters, 150, 247
  connectors, factory, 150, 247
  installation, 146, 246—247
  motorized, 246
  windshield, 246
ATC fuses and fuse holders, 166—168
ATO fuses and fuse holders, 166—168
ATRAC (Adaptive Transform Acoustic Coding), 126
Aura Bass Shakers, 119—120

automatic turn-on wiring, amplifiers, 168—169
auto reverse, cassette head units, 135

B

backstraps, 141
balanced line driver outputs, head units, 130
balanced signals, 255—256
banana plugs, 108—109
bandpass box:
   advantages of, 73—75
   box design software, 82—83
   construction, 104
   damping, enclosure, 111—112
   design example, 96—98
   one-note thump, 93
   port design example, 106—107
   port length, calculating, 105—106
   ports, constructing, 105—106
   ports, multiple, 107—108
   port tuning, 93—94
bass blockers, 114
bass blocking crossovers, 114—116
bass engines, 119—120
bass response goal, 92
bass shakers, 119—120
bass transducers, 119—120
battery monitors, 240—241
battery savers, 240—241
biamping:
   adjusting your system, 219—223
   advantages of, 206—208
   amplifiers for, 210—211
   choosing a crossover, 208—210
   configurations for head units with one set of
      preamp outputs, 215—219
   configurations for head units with two sets of
      preamp outputs, 211—215
   speakers for, 211
binding posts, 108—109
books, recommended, 280
Bose (*see* premium factory sound systems)
box design, 82—102
   bandpass box construction, 104
   bandpass design example, 96—98
   box assembly and bracing, 103—104
   box design charts, 85—91

box design (*Cont.*):
   box materials, 102—103
   box shape, 102—103
   construction, 102—113
   damping, enclosure, 111—112
   finishing touches, 112—113
   future of, 98
   isobarik configurations, 99—102
   multiple drivers in one box, 98—102
   port, when to use, 92
   port design example, 106—107
   port length, calculating, 105—106
   ports, constructing, 105—106
   ports, multiple, 107—108
   sealed and ported design example, 94—96
   software, 82—83
   terminals, speaker, 108—109
bridging, 67—68, 159—161

C

cables (*see* patch cables)
cable ties, 18
capacitors, power line, 271—272
carpet, speaker, 112
cassette adapters, for portable CD players, 244
cassette features, head units, 135—137
cassette head units, 125—126
CD changer controls, head units, 129
CD changers:
   for aftermarket head units with CD changer
      controls, 229
   connecting, 234—235
   converters, 230
   disc naming, 231
   ESP (Electronic Shock Protection), 232
   features, 230—232
   FM modulator, 227—228, 231
   for head units without CD changer controls,
      226—229, 230
   installation, 232—235
   with integral amplifier, 228—229
   mounting, 234
   mounting angle, 231
   mounting location, 232—234
   performance, 230—232
   shuffle play, 232

CD changers (*Cont.*):
   track programming, 231
   zero-bit detect, 232
   (*See also* head units, CD changer built-in)
CD features, head units, 137—139
CD head units, 125—126
characteristic impedance, of patch cables, 243
chokes, power line, 272
Chrysler head units, 122—123, 142—144
Chrysler/Infinity (*see* premium factory sound
     systems)
circuit breakers, power wiring, 167
claw tool, 24
clipping, 207—208
component speakers (*see* separates)
computer programs for box design, 82—83
connectors:
   crimp, 11—12
   power wiring, 170—171
   Scotchlok, 12—14
   wire nuts, 14
converters (*See also* adapters)
   CD changer, 230
copper shielding tape, 275—276
crimp connectors, 11—12
crimper/stripper, 19
crossovers:
   for biamping, 208—210
   capacitors for tweeters, 39—40
   cutoff frequency for subs, 64
   filter slope for subs, 63
   noise immunity considerations, 260
   passive, 203
   preamp-level, 202—204
   purpose, 202
   setting frequency, 219—220
   slope, 204—206, 208—209
   speaker-level, 202—204
   tri-way, 58—59
current, water analogy for, 163

D

damping, enclosure, 111—112
damping factor, 171
dashboard controls, factory, 151
Delco/Bose (*see* Premium factory sound systems)

differential signals, 255—256
DIN head units, 122—124, 144—145
DIN removal tool, 26
distortion, 266—267
distribution blocks, power wiring, 165—166
diversity tuning, 132
Dolby noise reduction, 135
double DIN head units, 122—123
dual voice coil speakers, 80—82
Dynamat, 248—249

E

efficiency (*see* sensitivity, speakers)
EMR sniffers, 273—275
equalizer boosters, 194
equalizers:
   bands, number needed, 192—194
   connecting, 194—195
   noise immunity considerations, 260
   parametric, 192—194
   setting by ear, 195—196
   setting by RTA, 196—199
ESP (Electronic Shock Protection), 137—138

F

f_{fb}, 77
f_{ob}, 79, 84
f_s (*see* T/S parameters)
factory look, 3
Faraday's law, 256
fiberglass damping, 111—112
filters, power line, 270—273
FM boosters, 247—248
FM modulator CD changers, 227—228
FM selectivity, 135, 248
FM sensitivity, 134, 248
FM transmitters, for portable CD players,
    244—245
foam speaker baffles, 44
Ford/JBL (*see* premium factory sound systems)
Ford Premium Sound (*see* premium factory
    sound systems)
free-air subwoofers, 72—73
Front Image Enhancer, 129

fuses, power wiring, 166—168
fuse taps:
    for AGC fuse blocks, 15
    for ATO/ATC fuse blocks, 15—16
    installation tips, 17—18
    lead-wire fuse, 15
    for MINI fuse blocks, 16

G

galvanized bodies, grounding to, 170
GM head units, 122—123, 142—144
ground, definition, 31
ground impedance, common, 257—258
grounding:
    to galvanized bodies, 170
    noise considerations, 264—266
ground loop isolators, 257, 269—270
ground loops, 256—257

H

head units:
    antenna mounting, 146
    auto reverse, cassette, 135
    backstraps, 141
    balanced line driver outputs, 130
    cassette, 125—126
    cassette features, 135—137
    cassette frequency response, 136
    CD, 125—126
    CD changer built-in, 137—138
    CD changer controls, 129, 229
    CD features, 137—139
    chassis-grounded speaker wiring, 149—150
    Chrysler, 122—123
    common-grounded speaker wiring, 149—150
    depth, mounting, 124
    detachable face, 128—129
    DIN, 122—123
    diversity tuning, 132
    Dolby noise reduction, 135
    double DIN, 122—123
    electrical wiring, 146—150
    ESP (Electronic Shock Protection), 137—138

head units (*Cont.*):
    features, 127—139
    FM selectivity, 135
    FM sensitivity, 134
    Front Image Enhancer, 129
    GM, 122—123
    high-power, 131
    ID-Logic, 132
    installation, 139—150
    ISO-DIN, 122—123
    ISO-DIN-J, 122—123
    MASK, 128
    mechanical mounting, 139—146
    MiniDisc, 125—126
    mounting kits, 139—141
    mounting methods, 122—123
    noise immunity considerations, 259
    OEM integration adapters, 148—154
    performance, 127—139
    preamp outputs, 130
    preamp output voltage, 130
    premium factory sound systems, 149, 151—154
    radio features, 132—135
    RDS (Radio Data System), 133—134
    remote control, 130
    RMS power, 130
    shuffle play, 138
    sizes, 122—123
    soft-touch controls, 137
    theft deterrents, 128—129
    Thummer remote control, 130—131
    universal, 122—123
    wiring harness adapters, 147—148
    zero-bit detect, 139
high-level to low-level converters (*see* line output converters)

I

ID-Logic, 132
imaging:
    definition, 34—35
    good, achieving, 36
in-dash receivers (*see* head units)
induced noise, 252

interconnects, 242—243
isobarik:
    configuration, 99—102
    driver mounting, 110—111
ISO-DIN head units, 122—123, 145—146
ISO-DIN-J head units, 122—123

K

kick panel pods, 53—54

L

lead-wire fuse, 15
line drivers, 261
line drop (*see* voltage drop)
line noise, 258—259
line output converters, 179
    premium factory sound systems, 62, 186
LOC (*see* line output converters)

M

magazines, recommended, 280—281
mail-order suppliers, 279
MASK, 128
MAXI fuses and fuse holders, 166—168
maximum power amplifier rating, 156—157
MDF (medium-density fiberboard), 102—103
MEGA fuses and fuse holders, 166—168
metal cutting tips, 49
MiniDisc, 125—126
multimeters, 26—28
    using, 28
multipath, 132
multitesters (*see* multimeters)
muting plugs, 268
MVC (master volume control), 261—263

N

noise:
    alternator whine, 258—259
    amplifiers, buying, 260

noise (*Cont.*):
    balanced signals, 255—256
    cable routing, 264
    capacitors, 271—272
    chokes, 272
    common ground impedance, 257—258
    crossovers, buying, 260
    differential signals, 255—256
    electronic suppressors, 273
    equalizers, buying, 260
    equipment placement, 262—264
    Faraday's law, 256
    filters, power line, 270—273
    grounding, 264—266
    ground loop isolators, 257, 269—270
    ground loops, 256—257
    head units, buying, 259
    high-power systems, 254—255
    high preamp levels, 253—254
    induced, 252
    line, 258—259
    line drivers, 261
    MVC (master volume control), 261—263
    patch cables, 260—261
    power supply, 258—259
    road (*see* road noise)
    shielding tape, 275—276
    shorting plugs, 268
    signal-to-noise ratio, 252—253
    sniffers, 273—275
    sources, identifying, 268—269
    system level adjustment, 266—267
    thermal, 252
    troubleshooting, 276—277
    twisted-pair cabling, 256
nutdrivers, 20

O

OEM integration adapters, 148—154
offset screwdriver, 19
Ohm's law, 163
oxygen-free copper, in patch cables, 243

P

paint pen, 25
parallel wiring, 66—69
parametric equalizers, 192—194
patch cables, 242—243
   noise immunity considerations, 260—261
   remote turn-on wire, 243
   routing to reduce noise, 264
   twisted-pair, 256
patch cords, 242—243
peak power amplifier rating, 156—157
pink noise, 197
ported box:
   advantages of, 73—74
   box design software, 82—83
   damping, enclosure, 111—112
   design example, 94—96
   port length, calculating, 105—106
   ports, constructing, 105—106
   ports, multiple, 107—108
   port tuning, 93—94
ports:
   constructing, 105—106
   length equation, 105—106
   multiple, 107—108
   when to use, 92
   (*See also* bandpass box; ported box)
power, water analogy for, 163
power line capacitors, 238—240
   ESL (equivalent series inductance), 239
   ESR (equivalent series resistance), 239
   installation, 240
   precharging, 240
   temperature rating, 239
power line filters, 270—273
power supply noise, 258—259
preamp outputs, head units, 130
preamp output voltage, head units, 130
premium factory sound systems, 149, 151—154
   basics, 152—153, 184—185
   boosting, 184—189
   bypassing factory amps, 185—186
   line output converters for, 186
   speakers, 185

Q

Q_{ts} (*see* T/S parameters)
quick splice connectors (*see* Scotchlok connectors)

R

radios (*see* head units)
RBDS (*see* RDS)
RDS (Radio Data System), 133—134
rear fill, 36
receivers (*see* head units)
relay for more current drive, 169
remote control, head units, 130
remote turn-on wiring, amplifiers, 168—169
   (*See also* patch cables, remote turn-on wire)
resistance, water analogy for, 163
RMS power:
   amplifier rating, 156—157
   head units, 130
road noise, 196
   masking, 196
   (*See also* sound-deadening materials)
RTA (real-time analyzer), 196—199

S

Scotchlok connectors, 12—14, 148
sealed box:
   advantages of, 72—74
   box design software, 82—83
   damping, enclosure, 111—112
   design example, 94—96
selectivity (*see* FM selectivity)
sensitivity, FM (*see* FM sensitivity)
sensitivity, speakers, 43, 76—77
separates (*see also* biamping), 52—53
series wiring, 66—69
shielding tape, 275—276
shorting plugs, 268
shuffle play, 138, 232
signal-to-noise ratio, 252—253
SnakeLight, 23
sniffers, noise, 273—275
SNR, 252—253

soft-touch controls, head units, 137
soldering, 10—11
  crimped connections, 170—171
  irons, 19
  tips for, 11
Sonotube, 103
sound-deadening materials, 248—249
soundproofing materials, 248—249
soundstage (*see* imaging)
speakers:
  adapters, 43—44
  for biamping, 211
  chassis-grounded (*see* OEM integration
      adapters)
  choosing, 41—44
  choosing a location, 45—47
  common-grounded (*see* OEM integration
      adapters)
  cone material, 43
  foam baffles, 44
  installing in door, 47—51
  kick panel pods, 53—54
  parallel wiring, 66—69
  polarity, 51
  polarity tester, 51
  premium factory sound systems, 185
  sensitivity, 43
  separates (component speakers), 52—53
  series wiring, 66—69
speaker wire, 171, 241—242
speed clips, 51—52
staging (*see* imaging)
steering wheel controls, factory, 151
Stiffening capacitors (*see* power line capacitors)
subsonic filter, 64
Subwoofer Design Toolbox, 83, 279
subwoofers:
  adding to premium factory sound systems, 62
  amplified, 61—62
  bandpass box construction, 104
  bandpass design example, 96—98
  bandpass enclosure type, 73—75
  bass response goal, 92
  bass response potential, 76—80
  box assembly and bracing, 103—104
  box construction, 102—113
  box design, 82—102
  box design charts, 85—91

subwoofers (*Cont.*):
  box design software, 82—83
  box materials, 102—103
  choosing a driver, 75—82
  choosing an amplifier for, 65—66
  comparing bass potential, 77—80
  cone material, 80
  configurations, 56—62
  crossover considerations, 63—65
  damping, enclosure, 111—112
  dual voice coil, 80—82, 159
  enclosure types, 71—75
  finishing touches, 112—113
  free-air, 72—73
  installation, 113—119
  parallel wiring, 66—69
  port design example, 106—107
  ported box enclosure type, 73—74
  port length, calculating, 105—106
  ports, constructing, 105—106
  ports, multiple, 107—108
  power handling, 75—76
  Q-Customs enclosure, 71
  ready-made, 70
  sealed and ported design example, 94—96
  sealed box enclosure type, 72—74
  secret to tight bass with, 118—119
  securing, 116—117
  sensitivity, 76—77
  series wiring, 66—69
  Stealthbox enclosure, 70
  system adjustment procedure, 118—119
  terminals, speaker, 108—109
  transmission line enclosure type, 73—75
  trunk sonic isolation, 116
  when to use a port, 92
  (*See also* bass transducers)
suppliers, mail-order, 279

T

T/S parameters:
  introduction to, 77—78
  Q_{ts}, high values, 79
terminals, speaker, 108—109
test CDs, 26—27
theft deterrents, head units, 128—129

thermal noise, 252
Thiele/Small parameters (*see* T/S parameters)
Thummer remote control, head units, 130—131
tie wraps (*see* cable ties)
tools:
    claw tool, 24
    crimper/stripper, 19
    DIN removal tool, 26
    multimeter, 26—28
    nutdrivers, 20
    offset screwdriver, 19
    paint pen, 25
    SnakeLight, 23
    soldering iron, 19
    test CDs, 26—27
    Torx drivers, 22
    utility hacksaw, 24
    utility knife, 25
    window clip remover, 25
Torx drivers, 22
transferability, when planning a system, 2
transmission line enclosure type, 73—75
tri-mode crossover (*see* tri-way crossover)
tri-way crossover, 58—59
troubleshooting noise problems, 276—277
tweeters:
    choosing, 37
    connecting, 39
    crossover caps, 39—40
    damage, by clipping, 207—208
    dome, 37
    level matching, 40—41
    piezo, 37
    placing, 38—39
    surface-mount, 36—37
twisted-pair cabling, 256
two-ohm stability, amplifiers, 159

U

universal head units, 122—123
upgradability, when planning a system, 3
upgrade solutions to common complaints, 5
utility hacksaw, 24
utility knife, 25

V

V_{as} (*see* T/S parameters)
V_{of}, 84
vented box (*see* ported box)
voltage, water analogy for, 163
voltage drop, 162—164, 170

W

window clip remover, 25
wire, speakers, 171, 241—242
wire gauge, amplifier wiring, 162—164
wire gauge, speaker wiring, 171
wire nuts, 14
wire ties (*see* cable ties)
wiring harness adapters, 147—148

X

Xmax, 76

Z

zero-bit detect, 139, 232

About the Author

Mark Rumreich has written articles for *Electronics Now, Speaker Builder, Radio-Electronics,* and *Audio Amateur* magazines and has taught seminars in loudspeaker design and automotive acoustics. In his capacity as a senior design engineer for Thomson Consumer Electronics, he specializes in analog and digital audio and video systems.